Geohydrology of Big Bear Valley, California: Phase 1—Geologic Framework, Recharge, and Preliminary Assessment of the Source and Age of Groundwater

Edited By Lorraine E. Flint and Peter Martin

With contributions by Justin Brandt, Allen H. Christensen, Alan L. Flint, Lorraine E. Flint, Joseph A. Hevesi, Robert Jachens, Justin T. Kulongoski, Peter Martin, and Michelle Sneed

Prepared in cooperation with Big Bear City Community Services District

Scientific Investigations Report 2012–5100

U.S. Department of the Interior
U.S. Geological Survey

U.S. Department of the Interior
KEN SALAZAR, Secretary

U.S. Geological Survey
Marcia K. McNutt, Director

U.S. Geological Survey, Reston, Virginia: 2012

For more information on the USGS—the Federal source for science about the Earth, its natural and living resources, natural hazards, and the environment, visit *http://www.usgs.gov* or call 1–888–ASK–USGS.

For an overview of USGS information products, including maps, imagery, and publications, visit *http://www.usgs.gov/pubprod*

To order this and other USGS information products, visit *http://store.usgs.gov*

Suggested citation:
Flint, L.E., and Martin, Peter, eds., with contributions by Brandt, Justin, Christensen, A.H., Flint, A.L., Flint, L.E., Hevesi, J.A., Jachens, Robert, Kulongoski, J.T., Martin, Peter, and Sneed, Michelle, 2012, Geohydrology of Big Bear Valley, California: Phase 1—Geologic Framework, Recharge, and Preliminary Assessment of the Source and Age of Groundwater: U.S. Geological Survey Scientific Investigations Report 2012–5100, 112 p.

Acknowledgments

This study was done in cooperation with Big Bear City Community Services District and Big Bear Lake Department of Water and Power. Their support and provision of data were invaluable for the completion of this study. Appreciation is extended to GeoScience Support Services, Inc. for supplying data and information as well.

Further appreciation is given to Greg Smith of the U.S. Geological Survey for assistance in the field and with data-quality tracking, and to Steve Crawford of the U.S. Geological Survey for well logging.

Contents

Figures

Tables

Conversion Factors

Inch/Pound to SI

Multiply	By	To obtain
Length		
inch (in.)	2.54	centimeter (cm)
inch (in.)	25.4	millimeter (mm)
foot (ft)	0.3048	meter (m)
mile (mi)	1.609	kilometer (km)
Area		
acre	4,047	square meter (m^2)
acre	0.4047	hectare (ha)
acre	0.4047	square hectometer (hm^2)
acre	0.004047	square kilometer (km^2)
square inch (in^2)	6.452	square centimeter (cm^2)
square mile (mi^2)	259.0	hectare (ha)
square mile (mi^2)	2.590	square kilometer (km^2)
Volume		
gallon (gal)	3.785	liter (L)
gallon (gal)	0.003785	cubic meter (m^3)
gallon (gal)	3.785	cubic decimeter (dm^3)
cubic inch (in^3)	16.39	cubic centimeter (cm^3)
cubic inch (in^3)	0.01639	cubic decimeter (dm^3)
cubic inch (in^3)	0.01639	liter (L)
cubic foot (ft^3)	28.32	cubic decimeter (dm^3)
cubic foot (ft^3)	0.02832	cubic meter (m^3)
acre-foot (acre-ft)	1,233	cubic meter (m^3)
acre-foot (acre-ft)	0.001233	cubic hectometer (hm^3)
Flow rate		
acre-foot per day (acre-ft/d)	0.01427	cubic meter per second (m^3/s)
acre-foot per year (acre-ft/yr)	1,233	cubic meter per year (m^3/yr)
acre-foot per year (acre-ft/yr)	0.001233	cubic hectometer per year (hm^3/yr)
foot per second (ft/s)	0.3048	meter per second (m/s)
foot per minute (ft/min)	0.3048	meter per minute (m/min)
foot per hour (ft/hr)	0.3048	meter per hour (m/hr)
foot per day (ft/d)	0.3048	meter per day (m/d)
foot per year (ft/yr)	0.3048	meter per year (m/yr)
cubic foot per second (ft^3/s)	0.02832	cubic meter per second (m^3/s)
cubic foot per second per square mile [(ft^3/s)/mi^2]	0.01093	cubic meter per second per square kilometer [(m^3/s)/km^2]
gallon per minute (gpm)	0.06309	liter per second (L/s)
inch per year (in/yr)	25.4	millimeter per year (mm/yr)
Mass		
ounce, avoirdupois (oz)	28.35	gram (g)
pound, avoirdupois (lb)	0.4536	kilogram (kg)
Radioactivity		
picocurie per liter (pCi/L)	0.037	becquerel per liter (Bq/L)
Hydraulic conductivity		
foot per day (ft/d)	0.3048	meter per day (m/d)

Temperature in degrees Fahrenheit (°F) may be converted to degrees Celsius (°C) as follows:

$$°C=(°F-32)/1.8$$

Vertical coordinate information is referenced to North American Vertical Datum of 1988 (NAVD 88).

Horizontal coordinate information is referenced to the North American Datum of 1983 (NAD 83).

Altitude, as used in this report, refers to distance above the vertical datum.

Concentrations of chemical constituents in water are given either in milligrams per liter (mg/L) or micrograms per liter (µg/L).

Acronyms and Abbreviations

AW	available water
BCM	Basin Characteristic Model
bls	below land surface
CIMIS	California Irrigation Management Information Systems
CSD	Community Services District
DEM	digital elevation model
DWP	Department of Water Power
EDNA	Elevation Derivatives for National Applications
EM	Electromagnetic
ENSO	El Nino Southern Oscillation
GIDS	gradient plus inverse distance squared
GPS	Global Positioning System
HSPF	Hydrologic Simulation Program–FORTRAN
InSAR	Interferometric Synthetic Aperture Radar
MUID	map-unit-identifier
MWD	Municipal Water District
NCDC	National Climatic Data Center
NHD	National Hydrography Dataset
NOAA	U.S. National Oceanographic and Atmospheric Administration
NSME	Nash-Sutcliffe Model Efficiency
NWS	National Weather Service
PAEE	Percent Average Estimation Error
PDO	Pacific Decadal Oscillation
PET	potential evapotranspiration
pmc	percent modern carbon
PRISM	Parameter Regression on Independent Slopes Model
PRMS	Precipitation-Runoff Modeling System
r^2	R-squared
RAWS	Remote Automated Weather Stations
RTK	Real Time Kinematic
SAR	Synthetic Aperture Radar
SBC	San Bernardino County
SGWZ	Shallow Groundwater Zone
STATSGO	State Soil Geographic Database
STP	Standard Temperature and Pressure
TU	Tritium Units
USGS	U.S. Geological Survey
VSMOW	Vienna Standard Mean Ocean Water

Well Numbering System

Wells are identified and numbered according to their location in the rectangular system for the subdivision of public lands. Identification consists of the township number, north or south; the range number, east or west; and the section number. Each section is divided into sixteen 40-acre tracts lettered consecutively (except I and O), beginning with "A" in the northeast corner of the section and progressing in a sinusoidal manner to "R" in the southeast corner. Within the 40-acre tract, wells are sequentially numbered in the order they are inventoried. The final letter refers to the base line and meridian. In California, there are three base lines and meridians; Humboldt (H), Mount Diablo (M), and San Bernardino (S). All wells in the study area are referenced to the San Bernardino base line and meridian (S) Well numbers consist of 15 characters and follow the format 002NE002N001E12M001s. In this report, well numbers are abbreviated and written 2N/1E-12M1. Wells in the same township and range are referred to only by their section designation, 12M1. The following diagram shows how the number for well 2N/1E-12M1 is derived.

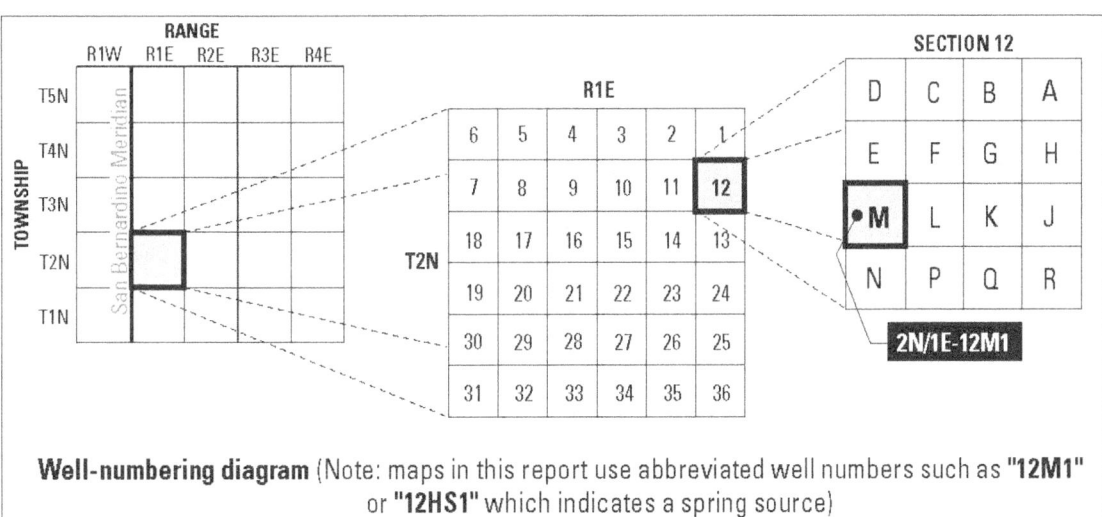

Well-numbering diagram (Note: maps in this report use abbreviated well numbers such as **"12M1"** or **"12HS1"** which indicates a spring source)

Geohydrology of Big Bear Valley, California: Phase 1—Geologic Framework, Recharge, and Preliminary Assessment of the Source and Age of Groundwater

Edited by Lorraine E. Flint and Peter Martin

Abstract

The Big Bear Valley, located in the San Bernardino Mountains of southern California, has increased in population in recent years. Most of the water supply for the area is pumped from the alluvial deposits that form the Big Bear Valley groundwater basin. This study was conducted to better understand the thickness and structure of the groundwater basin in order to estimate the quantity and distribution of natural recharge to Big Bear Valley.

A gravity survey was used to estimate the thickness of the alluvial deposits that form the Big Bear Valley groundwater basin. This determined that the alluvial deposits reach a maximum thickness of 1,500 to 2,000 feet beneath the center of Big Bear Lake and the area between Big Bear and Baldwin Lakes, and decrease to less than 500 feet thick beneath the eastern end of Big Bear Lake. Interferometric Synthetic Aperture Radar (InSAR) was used to measure pumping-induced land subsidence and to locate structures, such as faults, that could affect groundwater movement. The measurements indicated small amounts of land deformation (uplift and subsidence) in the area between Big Bear Lake and Baldwin Lake, the area near the city of Big Bear Lake, and the area near Sugarloaf, California. Both the gravity and InSAR measurements indicated the possible presence of subsurface faults in subbasins between Big Bear and Baldwin Lakes, but additional data are required for confirmation.

The distribution and quantity of groundwater recharge in the area were evaluated by using a regional water-balance model (Basin Characterization Model, or BCM) and a daily rainfall-runoff model (INFILv3). The BCM calculated spatially distributed potential recharge in the study area of approximately 12,700 acre-feet per year (acre-ft/yr) of potential in-place recharge and 30,800 acre-ft/yr of potential runoff. Using the assumption that only 10 percent of the runoff becomes recharge, this approach indicated there is approximately 15,800 acre-ft/yr of total recharge in Big Bear Valley.

The INFILv3 model was modified for this study to include a perched zone beneath the root zone to better simulate lateral seepage and recharge in the shallow subsurface in mountainous terrain. The climate input used in the INFILv3 model was developed by using daily climate data from 84 National Climatic Data Center stations and published Parameter Regression on Independent Slopes Model (PRISM) average monthly precipitation maps to match the drier average monthly precipitation measured in the Baldwin Lake drainage basin. This model resulted in a good representation of localized rain-shadow effects and calibrated well to measured lake volumes at Big Bear and Baldwin Lakes. The simulated average annual recharge was about 5,480 acre-ft/yr in the Big Bear study area, with about 2,800 acre-ft/yr in the Big Bear Lake surface-water drainage basin and about 2,680 acre-ft/yr in the Baldwin Lake surface-water drainage basin.

One spring and eight wells were sampled and analyzed for chemical and isotopic data in 2005 and 2006 to determine if isotopic techniques could be used to assess the sources and ages of groundwater in the Big Bear Valley. This approach showed that the predominant source of recharge to the Big Bear Valley is winter precipitation falling on the surrounding mountains. The tritium and uncorrected carbon-14 ages of samples collected from wells for this study indicated that the groundwater basin contains water of different ages, ranging from modern to about 17,200-years old.

The results of these investigations provide an understanding of the lateral and vertical extent of the groundwater basin, the spatial distribution of groundwater recharge, the processes responsible for the recharge, and the source and age of groundwater in the groundwater basin. Although the studies do not provide an understanding of the detailed water-bearing properties necessary to determine the groundwater availability of the basin, they do provide a framework for the future development of a groundwater model that would help to improve the understanding of the potential hydrologic effects of water-management alternatives in Big Bear Valley.

Introduction

The Big Bear Valley is an east-west trending valley that extends from the west end of Big Bear Lake to the east end of Baldwin Lake in the San Bernardino Mountains of southern San Bernardino County, California (*fig. 1*). The valley is a popular recreational area for residents in southern California. The water supply for the Big Bear Valley is produced from springs and slant wells in the mountains that surround the valley and from wells drilled within the groundwater basin that underlies the valley. Water use in the Big Bear area has increased significantly since 1980 as the permanent population of the valley increased from about 12,000 people in 1980 to about 22,000 people in 2004. In addition to the permanent residents, recreational visitors cause the population to swell to over 100,000 people on occasions during the summer and winter tourist seasons. A drought, which extended from 1998 through 2003, resulted in less recharge than average to the aquifer system and caused spring flow and groundwater levels to decline. In June 2003, the city of Big Bear Lake Department of Water and Power (DWP) Board of Commissioners declared a water-shortage emergency for the DWP service area in the Big Bear Valley. The Big Bear City Community Services District (CSD) limited new water connections and initiated water restrictions in response to the drought. To help meet water demand, the local water agencies constructed new wells and have studied artificial recharge with reclaimed wastewater. To better manage the groundwater resources in the Big Bear Valley, there is a need to better understand the geohydrology of Big Bear Valley and the surrounding area, in particular the size and shape of the groundwater basin, the quantity and distribution of natural groundwater recharge to the entire Big Bear Valley, and the effect of climate variability on natural groundwater recharge.

Purpose and Scope

The Big Bear City Community Services District (CSD) is interested in developing a better understanding of the geohydrology of the Big Bear Valley groundwater basin to help the district and other local agencies better manage and utilize the groundwater resources of the valley. Questions that need to be answered are the following: How large is the groundwater basin? What is the quantity of groundwater in storage? Are there buried geologic structures that could affect groundwater movement and water quality? What is the annual amount of groundwater recharge, and how will future climate variations affect the quantity of recharge? What are the source and age of groundwater in the basin? What is the spatial distribution of groundwater quality in the basin? How will future pumping and artificial recharge operations affect water levels and water quality? Because of limitations of funding, the CSD would like to answer these questions in phases, with the initial phase addressing the characterization of the groundwater basin. Future studies may be conducted, such as groundwater

modeling, which would further refine the characterization and quantify the groundwater in order to provide a means to address management questions regarding future pumping or amelioration operations.

The objectives of the first phase of the study are to (1) define the thickness and structure of the Big Bear Valley groundwater basin, (2) estimate the quantity, spatial and temporal distribution, and source of natural recharge to the groundwater basin, and (3) determine if isotopic techniques could be used to assess the source age of groundwater in the basin. The thickness and structure of the groundwater basin were investigated for this study by using gravity, aeromagnetic, and Interferometric Synthetic Aperture Radar (InSAR) data. Variation in the expression, created by analysis of InSAR data, of pumping-induced subsidence can be used to identify subsurface barriers to groundwater flow. The quantity, distribution, and source of groundwater recharge were estimated by using two numerical modeling approaches: (1) a regional-scale basin recharge model and (2) a detailed precipitation-runoff model. Selected springs and wells were sampled and analyzed to derive chemical and isotopic data to determine if there are significant variations in the source and age of water in the basin. Future phases of the study could include defining the groundwater quality of the basin and developing and calibrating a groundwater-flow model to simulate the groundwater-flow system and improve understanding of the spatial and quantitative processes. The objectives, scope, and timing of future phases of the study will be determined in consultation with CSD after evaluating the results of the first phase. This report is intended to document the described activities of this study, providing all methods, interpretations, uncertainties, and appropriate applications.

Description of the Study Area

Hydrography

Big Bear Valley groundwater basin (as defined by Bulletin 118, California Department of Water Resources, 2003, p. 88) is located at the base of a north-facing slope of the San Bernardino Mountains and extends from Big Bear Lake on the west to Baldwin Lake on the east (*fig. 1*). The groundwater basin lies within the Big Bear Lake and Baldwin Lake surface-water drainage basins (*fig. 1*). The groundwater basin occupies about 41 percent, or 49 square miles (mi^2) of the total area of 120 mi^2 for the surface-water drainage basins (). The surrounding mountains rise to approximately 7,800 to 10,200 feet (ft) above sea level along the ridge south of the groundwater basin. The area overlying the groundwater basin receives surface-water runoff from the Big Bear Lake and the Baldwin Lake surface-water drainage basins. The Big Bear Lake surface-water drainage basin includes seven subbasins defined by surface-water drainage divides: Gray's Landing, Grout Creek, North Shore, Division, Rathbone, Village, and Mill Creek. The Baldwin Lake surface-water drainage basin includes four subbasins defined by surface-water drainage

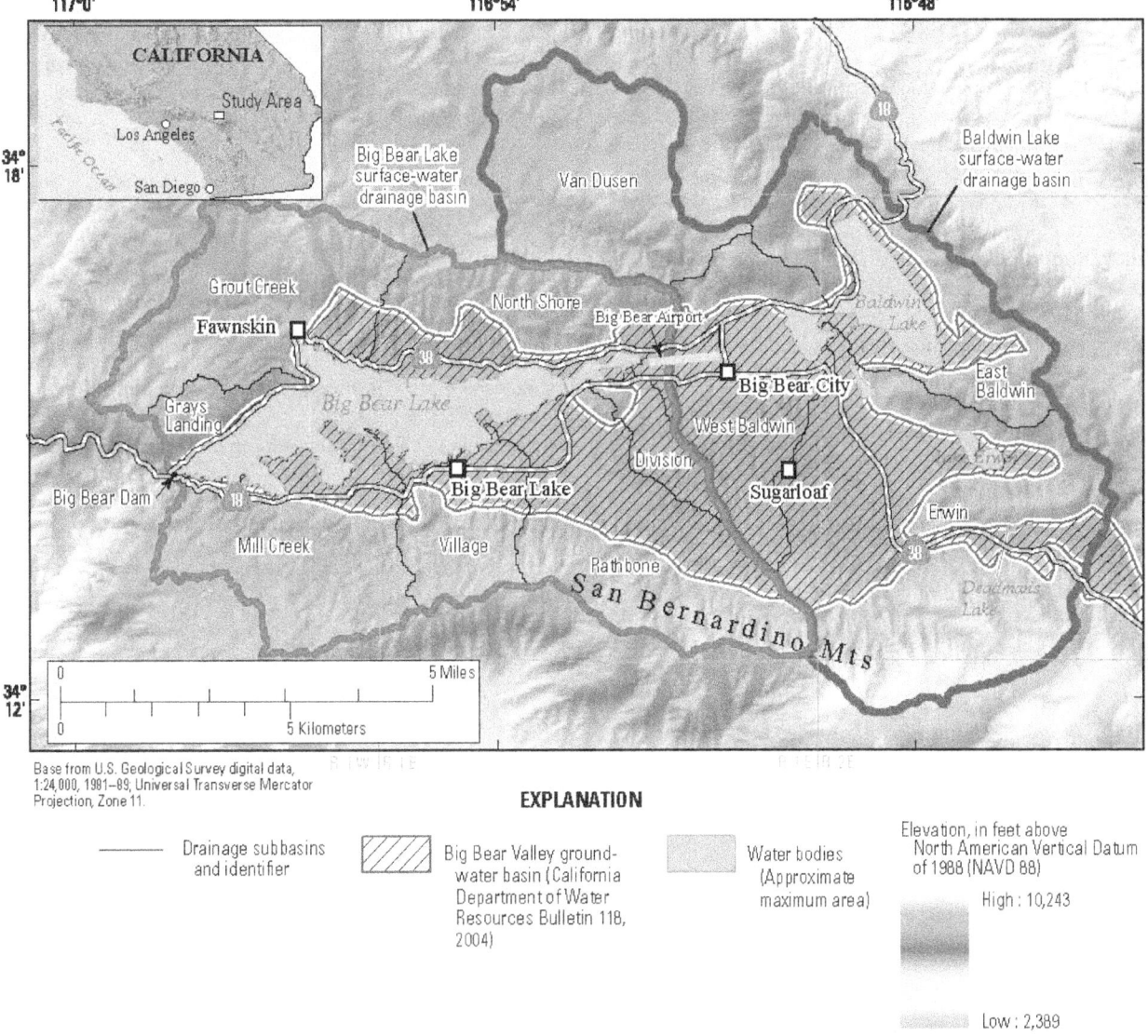

Figure 1. Location of study area, surface-water drainage basins, and groundwater basin of Big Bear Valley, San Bernardino County, California.

divides: Van Dusen, East Baldwin, West Baldwin, and Erwin. For the purposes of this report, the combined area of the Big Bear Lake and Baldwin Lake surface-water drainage basins is referred to as Big Bear Valley.

Big Bear Lake is a man-made lake that lies at an elevation of about 6,800 ft and is fed by runoff from numerous creeks that drain the mountains and valley floor. Big Bear Lake is contained by Bear Valley Dam at the west end of the lake. Baldwin Lake, typically dry, lies at an elevation of about 6,700 ft and receives occasional runoff from canyons in Van Dusen subbasin to the northwest and creeks in Erwin subbasin to the south. There are several other small natural lakes in the Baldwin Lake surface-water drainage basin, including

Lake Erwin (*fig. 1*), two small mountain lakes to the southeast of Lake Erwin, and an intermittent, marshy lake just south of Baldwin Lake. Only Big Bear Lake, Baldwin Lake, and Lake Erwin are considered significant hydrologic entities in the model development of this area. It is not known if there are subsurface connections among the lakes in the Baldwin Lake surface-water drainage basin, and the only significant surface-water outflow from any of the lakes is from Big Bear Lake to the west through Bear Valley Dam and into Bear Creek, which flows into the Santa Ana River 8 miles (mi) to the south.

Climate

Big Bear Valley has warm summers and cold winters, which is typical of mountainous areas in southern California. Average daily temperatures range from about 60 to 70 degrees Fahrenheit (°F) in the summer and 35 to 40°F in the winter (Crippen, 1965). Average annual precipitation in the valley ranges from about 35 inches (in.) on the western edge of the valley and in the mountains south of Baldwin Lake to 18 in. on the eastern edge of the valley (fig. 2). The precipitation distribution reflects a rain-shadow effect that is reflected in the distribution of vegetation, which ranges from montane hardwoods and mixed conifers in the mountains to the south and west, to pinyon-juniper woodlands and sagebrush around Baldwin Lake. Big Bear Valley has a unique ecosystem because of its latitude, elevation, and rainfall, which provides the habitat for several plant species, which are found nowhere else in California. The local climate, soils, geology, and topography conditions create a "pebble plain" habitat of treeless islands where vegetation includes low-growing perennial plants. Vernal meadow habitat also remains in a few locations throughout the area.

Geology

Big Bear Valley is in the geologically complex San Bernardino Mountains (fig. 3). For the purposes of this report, the geologic units are generalized into pre-Tertiary basement rocks, generally located in the mountains surrounding the valley floor, and alluvial deposits that form the Big Bear Valley groundwater basin. The basement rocks are dominated by (1) large Cretaceous granitic bodies ranging in composition from monzogranite to gabbro, (2) metamorphosed sedimentary

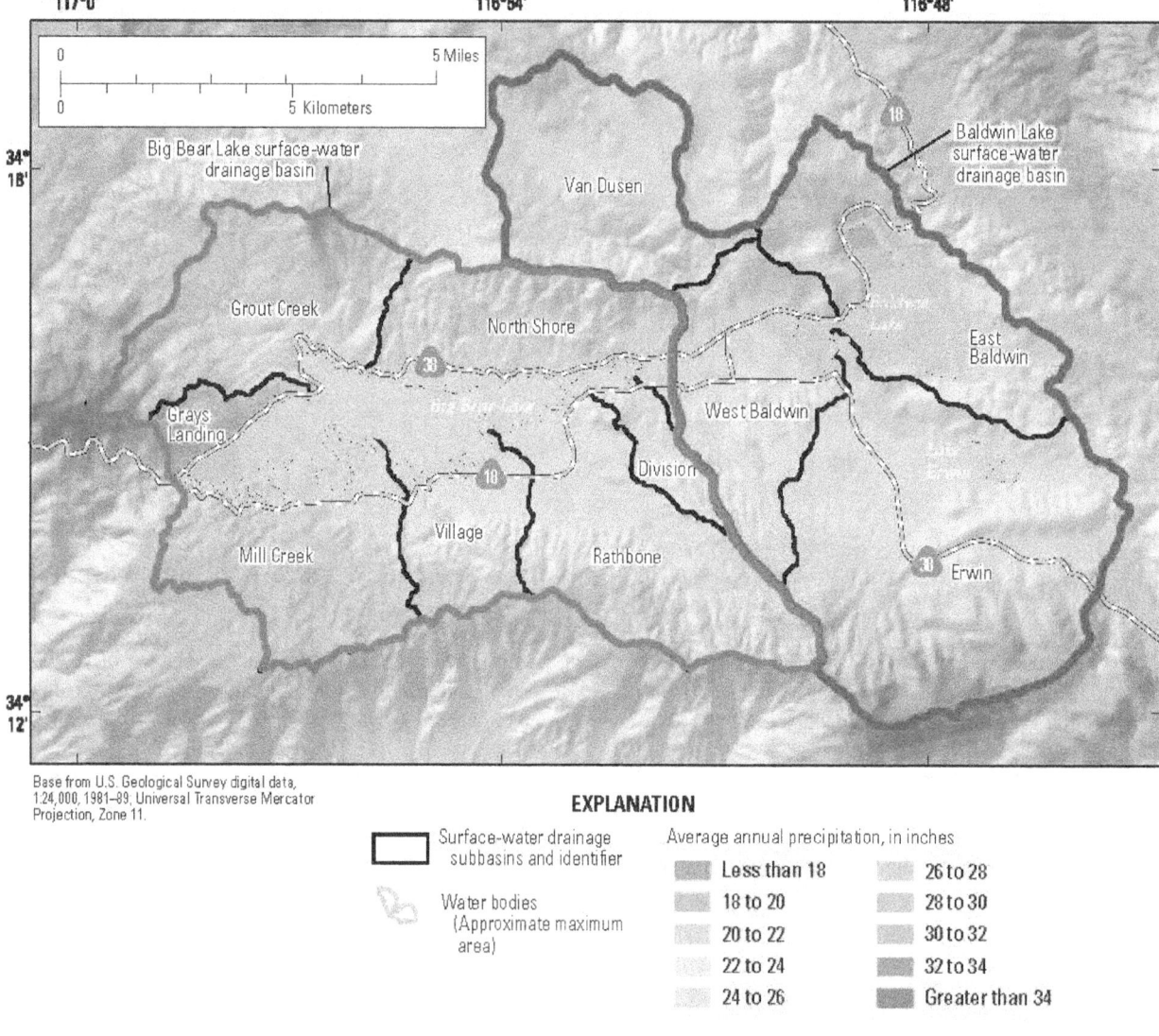

Base from U.S. Geological Survey digital data,
1:24,000, 1981–89; Universal Transverse Mercator
Projection, Zone 11.

EXPLANATION

☐ Surface-water drainage subbasins and identifier

〰 Water bodies (Approximate maximum area)

Average annual precipitation, in inches

Less than 18	26 to 28
18 to 20	28 to 30
20 to 22	30 to 32
22 to 24	32 to 34
24 to 26	Greater than 34

Figure 2.   Average annual precipitation of Big Bear Valley, San Bernardino County, California, 1971–2000.

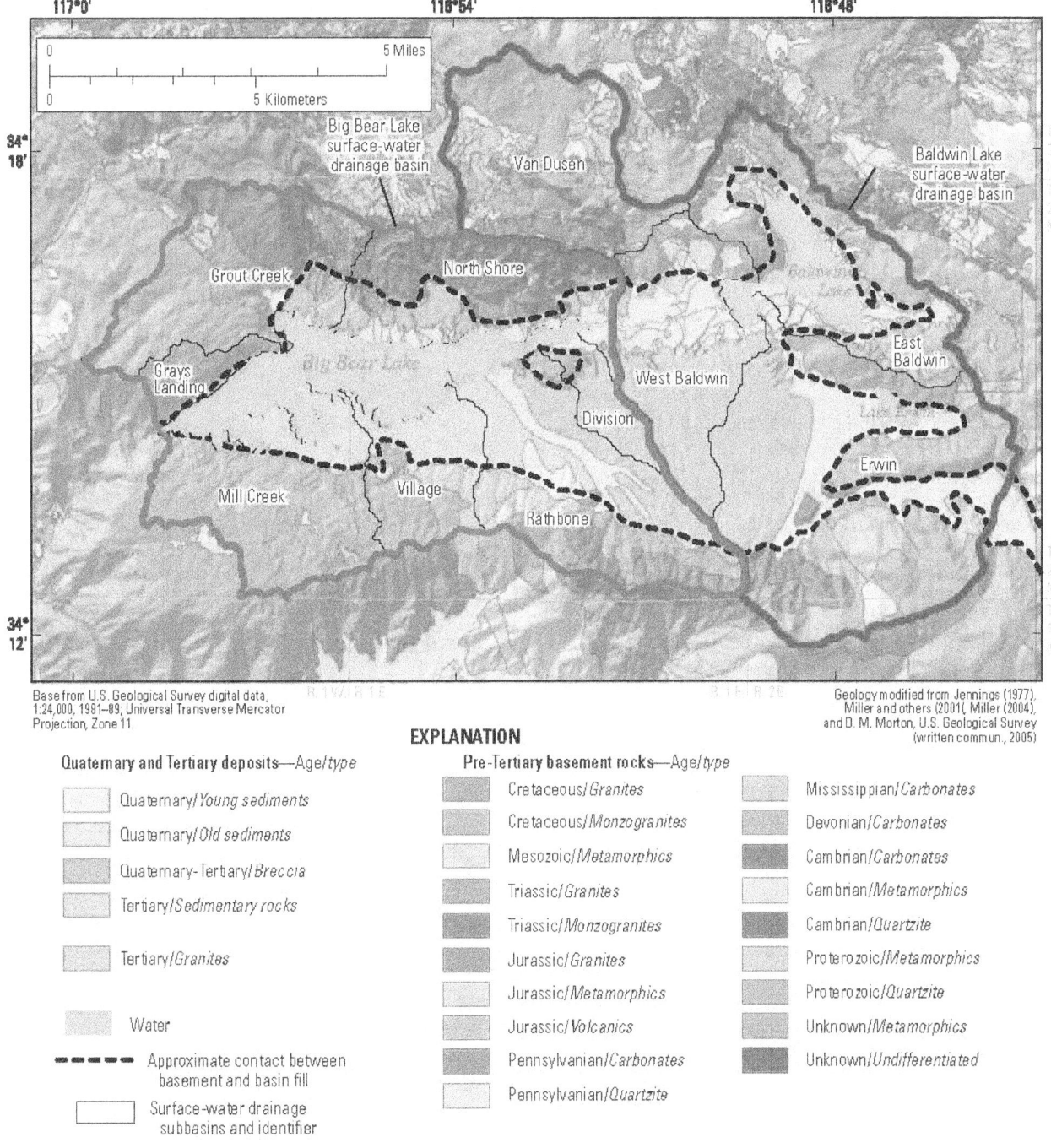

Base from U.S. Geological Survey digital data,
1:24,000, 1981–89; Universal Transverse Mercator
Projection, Zone 11.

Geology modified from Jennings (1977),
Miller and others (2001, Miller (2004),
and D. M. Morton, U.S. Geological Survey
(written commun., 2005)

EXPLANATION

Quaternary and Tertiary deposits—Age/*type*

Quaternary/*Young sediments*

Quaternary/*Old sediments*

Quaternary-Tertiary/*Breccia*

Tertiary/*Sedimentary rocks*

Tertiary/*Granites*

Water

- - - - Approximate contact between
basement and basin fill

Surface-water drainage
subbasins and identifier

Pre-Tertiary basement rocks—Age/*type*

Cretaceous/*Granites*

Cretaceous/*Monzogranites*

Mesozoic/*Metamorphics*

Triassic/*Granites*

Triassic/*Monzogranites*

Jurassic/*Granites*

Jurassic/*Metamorphics*

Jurassic/*Volcanics*

Pennsylvanian/*Carbonates*

Pennsylvanian/*Quartzite*

Mississippian/*Carbonates*

Devonian/*Carbonates*

Cambrian/*Carbonates*

Cambrian/*Metamorphics*

Cambrian/*Quartzite*

Proterozoic/*Metamorphics*

Proterozoic/*Quartzite*

Unknown/*Metamorphics*

Unknown/*Undifferentiated*

Figure 3.   General geology of Big Bear Valley, San Bernardino County, California.

rocks ranging in age from late Paleozoic to late Proterozoic, and (3) Middle Proterozoic gneiss (Miller, 2004). These rocks are complexly deformed by normal, reverse, and thrust faults, and are tightly folded in some places (Miller, 2004). In general, the basement rocks are of low permeability and are not considered a major water-bearing unit except in fractures and weathered zones that can create shallow perched groundwater zones.

Tertiary sedimentary deposits overlie basement rocks throughout most of the groundwater basin. This stratigraphic unit consists primarily of consolidated to partly consolidated alluvial-fan deposits and probably yields only small quantities of water to wells. These deposits are predominantly exposed on the southwest side of Baldwin Lake at the base of the mountains (*fig. 3*) and are reported to be greater than 1,000 ft thick in some areas (GeoScience Support Services, Inc., 2003).

Quaternary alluvial deposits overlie the Tertiary sedimentary rocks and basement rocks throughout much of the groundwater basin (*fig. 3*). For the purposes of this report, the Quaternary alluvial deposits shown on figure 3 were generalized into older alluvium and recent alluvium. The older alluvium consists predominantly of clay and sandy clay with interbedded layers of sand and gravel near Big Bear Lake and coarsens to predominantly sand with some gravel and interbedded layers of silt and clay toward Baldwin Lake (GeoScience Support Services, Inc., 1999). Beneath Baldwin Lake, the alluvial deposits consist of lacustrine deposits composed of clay and silt and interbedded sand (GeoScience Support Services, Inc., 2003). The coarse-grained layers in the older alluvium are probably the major water-bearing units in the groundwater basin.

Recent alluvium is present predominantly in the surface drainages in the watershed outside of the groundwater basin and in the shallow subsurface between Big Bear and Baldwin Lakes. The recent alluvium consists of permeable sand and gravel deposits with minor interbedded layers of silt and clay. Most of the recent alluvium is above the water table. Where present, the permeable recent alluvium allows rapid infiltration of available rainfall and runoff.

Groundwater Basin

The Big Bear Valley groundwater basin is surrounded and underlain by pre-Tertiary basement rocks of the San Bernardino Mountains. Precipitation in the surrounding and overlying surface-water drainage basins, as rain or snow, provides the water available for recharge to the underlying groundwater basin. Most of the water supply for the Big Bear area is pumped from the unconsolidated Quaternary alluvial deposits in the groundwater basin. The water-bearing deposits in the groundwater basin have been classified into upper, middle, and lower aquifers (GeoScience Support Services, Inc., 1999), and the upper and middle aquifers are the primary water producers. Inspection of geologic logs from wells drilled in the Baldwin Lake area indicates that the deposits consist of sands and gravels with interbedded clays.

Groundwater levels within the groundwater basin vary in response to long-term precipitation trends. In general, when precipitation amounts are in excess of average values, water levels rise, and when precipitation amounts are less-than-average values, water levels decline. In addition to natural variations, water levels respond to changes in groundwater pumping. Long-term pumping in excess of natural recharge will result in water-level declines until the groundwater system reaches a new equilibrium by removing water from storage, decreasing natural discharge, increasing recharge to the system, or some combination of these processes. Natural variability, such as drought, can require re-establishment of equilibrium conditions as water levels decline as a result of decreased recharge and increased pumping (Alley and others, 1999). Most groundwater development is very complex and can comprise many wells pumping from an aquifer at varying pumping rates and at different locations within the groundwater-flow system. Computer models commonly are needed to evaluate the temporal response of groundwater levels to complex patterns of groundwater development (Alley and others, 1999).

Groundwater within the basin generally has low concentrations of dissolved solids; however, groundwater pumped from some wells in the Baldwin Lake area has fluoride concentrations in excess of 6.0 milligrams per liter (GeoScience Support Services, Inc., 2001). A comprehensive assessment of groundwater quality in the Big Bear Valley has not been completed.

History of Big Bear Lake

Big Bear Valley has a rich and varied history mostly associated with the natural and man-made lakes. Originally, Big Bear Valley was called Yuhaviat after a Serrano Indian word that means "Pine Place," and it was the name the area had for more than a thousand years (*http://www.bigbear.com/about/history/*). However, in 1845, Benjamin Davis Wilson rode into Yuhaviat Valley with several cowboys from the nearby town of Riverside. As they entered the valley, they found it swarming with bears. Wilson divided his men into two-man teams, each of whom went out, roped a bear, and brought it back to camp. With eleven bears at the camp at one time, Wilson came up with the name Big Bear Lake. However, it should be noted that Big Bear Lake is a man-made lake that didn't exist in 1845. The lake Wilson named Big Bear was actually the natural lake at the east end of the valley, now known as Baldwin Lake.

The original Bear Valley Dam (Old Bear Valley Dam) was built in 1884 to create a reservoir to meet the irrigation needs of the downstream growers, primarily citrus farmers. The Bear Valley Mutual Water Company was formed in 1903 by the growers, and they took over the operation of the dam and the lake in 1909. The present multiple-arch dam (New Bear Valley Dam) was constructed between 1910 and 1912 about 300 ft downstream from the Old Bear Valley Dam. In the late 1950s and early 60s, the southern California area was in the midst of a long and extensive drought. Because of extremely large demands on water from Big Bear Lake for irrigation in the San Bernardino/Redlands area, the lake was reduced to little more than a large "mud puddle" (fig. 4).

In 1964, the community of Big Bear Lake decided to gain control of Big Bear Lake for recreational purposes. The main difficulty lay in the fact that the water level of an irriga- tion reservoir, by its nature, changes drastically to meet the irrigation needs downstream; whereas, recreational interests required a reasonably stable water level. In 1964, by an over- whelming vote, the people of Big Bear Lake created the Big Bear Municipal Water District (MWD) with the express purpose of attempting to stabilize the level of Big Bear Lake. In 1977, following a long legal battle, the MWD acquired title to the dam, the area lying beneath the lake, and the surface-recreation rights to Big Bear Lake.

With regard to the water rights, all parties to the original lawsuit agreed to a stipulated judgment in the adjudication of the water rights, establishing a physical solution to the water rights dispute. The physical solution is a method whereby the MWD can maintain water in the lake while, at the same time, the irrigation interests downstream can be satisfied. In prac- tice, each year Bear Valley Mutual Water Company determines the irrigation needs downstream and estimates the demand on Big Bear Lake to meet these needs. The MWD then has the option of either supplying this needed water from another source (mainly the State Water Project and the Upper Santa Ana groundwater basin) or releasing the water from the lake. In this manner, the MWD can maintain the lake at an eleva- tion significantly above its pre-1964 levels. Over the years, the MWD has implemented several management strategies to maintain the level of the lake in the most cost-effective man- ner possible, and a 1996 water-purchase agreement with San Bernardino Valley Municipal Water District (Valley District) has helped MWD achieve its mission of lake stabilization. Although stabilizing the lake level provides some assurance of water supply during severe climatic variation, the stresses on the groundwater supply from increased population and the potential long-term reduction of water supply as a result of climate change require a more long-term approach and an attempt to understand the geohydrology and better manage the groundwater resources in addition to the surface-water supply.

Figure 4.   Photo of Big Bear Lake (looking east), in 1956 during a series of droughts, Big Bear Valley, San Bernardino County, California.

Thickness and Structure of the Big Bear Valley Groundwater Basin

The thickness and structure of the Big Bear Valley groundwater basin were investigated for this study using gravity, aeromagnetic, and Interferometric Synthetic Aperture Radar (InSAR) data. A description of the data, methods of data collection and interpretation, and results are presented in this report.

Gravity Survey

By Robert Jachens and Allen H. Christensen

A gravity survey was used to estimate the thickness of the alluvial deposits (Quaternary alluvial deposits and Tertiary sedimentary deposits) that form the Big Bear Valley ground-water basin and to understand the three-dimensional structure

(geometry) of the groundwater basin. Gravity measurements and water-well logs were the primary data sets used to define the thickness and structure of the groundwater basin. There is a large contrast between the densities of the alluvial deposits and the basement rocks, which averages 300 kilograms per cubic meter (kg/m^3), making the depth to basement rocks a good target for study by gravity methods. Aeromagnetic data also were used to help define the nature of the basement rocks that lie beneath the groundwater basin. Geologic maps were used to define the contact of the basement rocks and alluvial deposits at land surface.

Gravity Data

Gravity measurements were made at 305 locations in the Big Bear Valley and combined with regional gravity data (Roberts and others, 1990) to produce an isostatic residual gravity-field map (fig. 5). Gravity was measured during this study by using a LaCoste and Romberg Model D-79 with

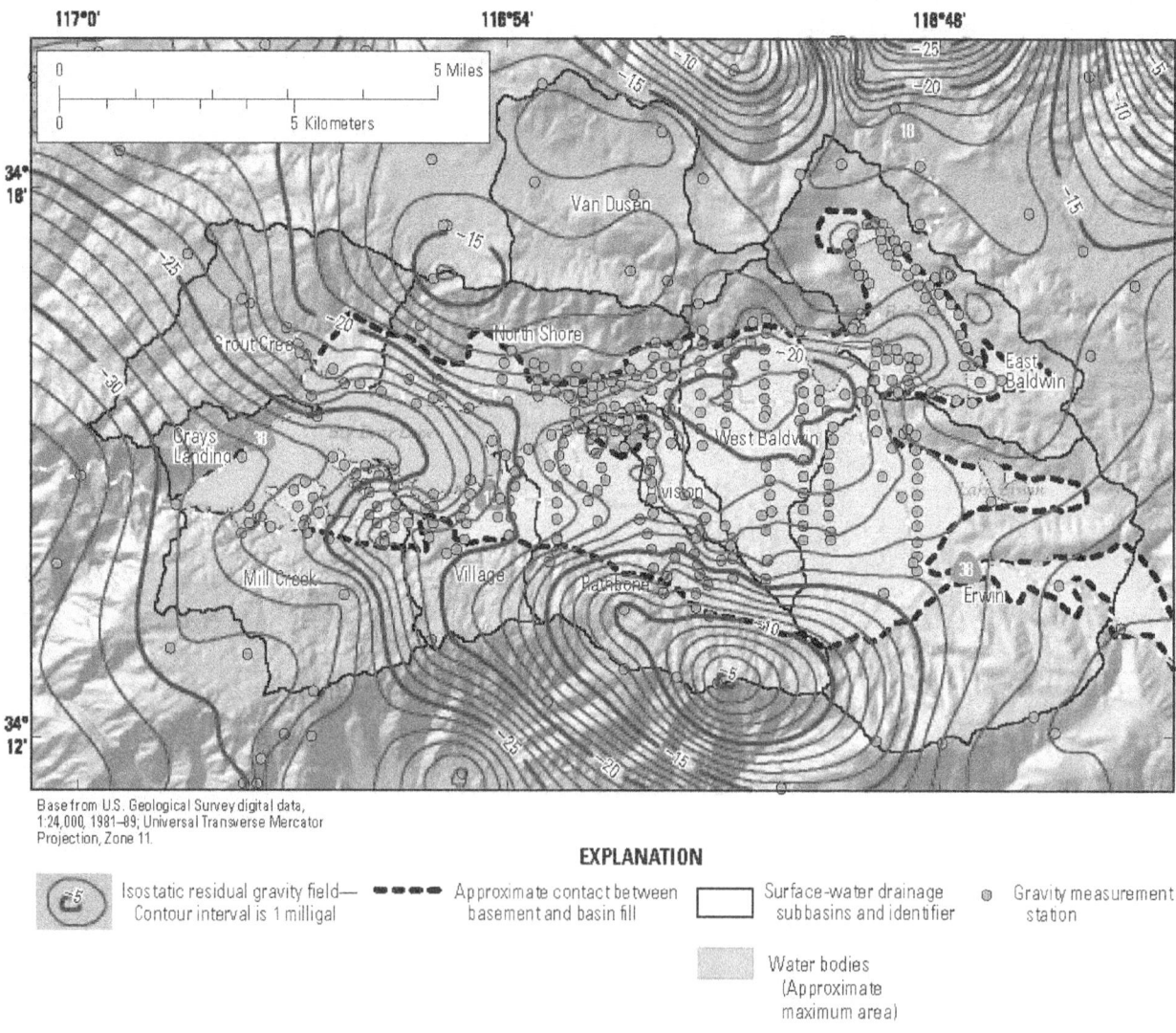

Base from U.S. Geological Survey digital data,
1:24,000, 1981–89; Universal Transverse Mercator
Projection, Zone 11.

EXPLANATION

○⁵ Isostatic residual gravity field— Contour interval is 1 milligal

▬▬ Approximate contact between basement and basin fill

☐ Surface-water drainage subbasins and identifier

⊚ Gravity measurement station

▨ Water bodies (Approximate maximum area)

Figure 5.   Gravity stations and measured isostatic gravity field of Big Bear Valley, San Bernardino County, California.

Aliod 100 gravity meter. Most of the measurements were made along north-south transects across Big Bear Valley groundwater basin; more than half of the measurements were made on the alluvial deposits that form the groundwater basin. The location and elevation of each gravity measurement was determined using a Trimble Real Time Kinematic (RTK) Model 4400 Global Positioning System (GPS) base and mobile receivers. This system can determine vertical and horizontal coordinates with a precision of plus or minus 0.083 ft between receiver and base.

Gravity data were analyzed using standard gravity corrections, including (a) the earth tide correction, which corrects for tidal effects of the moon and sun; (b) instrument drift correction, which compensates for drift in the instrument's spring; (c) the latitude correction, which incorporates the variation of the Earth's gravity with latitude; (d) the free-air correction, which accounts for the variation in gravity caused by a variation in elevation relative to sea level; (e) the Bouguer correction, which corrects for the attraction of material between the station and sea level; (f) the curvature correction, which corrects the Bouguer correction for the effect of the Earth's curvature; (g) the terrain correction, which removes the effect of topography to a radial distance of about 104 mi; and (h) the isostatic correction, which removes long-wavelength variations in the gravity field inversely related to topography.

Terrain corrections involved a 3-part process: (1) Hayford-Bowie zones A and B with an outer radius of about 220 ft were estimated in the field with the aid of tables and charts, (2) Hayford-Bowie zones C and D with an outer radius of about 1,940 ft were computed using a 30-ft, or 10-meter (m), digital elevation model, and (3) terrain corrections from a distance of about 1,940 ft to 104 mi were calculated by using a digital elevation model and a procedure by Plouff (1992). The regional isostatic gravity field was removed from the Bouguer field by using an Airy-Heiskanen model for isostatic compensation of topographic loads (Jachens and Roberts, 1981) that assumed a crustal thickness of about 15.5 mi, a crustal density of 2.67 grams per cubic centimeter (g/cm³), and a density contrast across the base of the model of 0.4 g/cm³, and that used topography averaged over 3 by 3 minute compartments to a distance of 104 mi from each station. Isostatic and terrain corrections beyond that distance were interpolated from a grid generated from Karki and others (1961).

Gravity Field

The gravity field of the study area (the isostatic residual gravity field) is complex and mostly reflects lateral variations of density of the basement rock (*fig. 5*). The most prominent anomaly on the gravity map is the elongate gravity high (−5 milligal in the center) on the southeastern edge of the Big Bear Lake surface-water drainage basin. This high is caused by a dense, mafic intrusive body of Jurassic hornblende diorite. Gravity intensity measured over the exposed basement rocks north of Big Bear Valley groundwater basin is substantially lower and reflects basement rocks having moderate densities. Gravity is lowest along the western edge of the study area because of a low-density intrusive body of Cretaceous age that lies mostly west of the study area.

Groundwater Basin Thickness Computation Method

The thickness of the alluvial deposits that form the groundwater basin was estimated by the method of Jachens and Moring (1990), modified slightly to permit inclusion of constraints at points where the thickness of the alluvial deposits is known from direct observations in drill holes. This method partitions the residual gravity field into two components—the component caused by density variations within the basement rocks (the basement gravity field) and the component caused by the low-density alluvial deposits that form the groundwater basin (the "groundwater basin gravity anomaly"). Once the gravity data have been partitioned, the groundwater basin gravity anomaly can be modeled to yield a thickness of the alluvial deposits throughout the study area, given the difference between the densities of the deposits that form the groundwater basin and the basement rocks.

An initial estimate of the groundwater basin gravity anomaly was made by passing a smooth surface through the gravity values at stations where the basement rocks are exposed at land surface (initial estimate of the "basement gravity field") and subtracting the result from the total gravity field (*fig. 5*). All gravity measurements made on mapped basement rocks were considered basement values. This smooth surface represented only the initial estimate because the gravity intensities measured on the exposed basement rocks close to the boundary of the groundwater basin are influenced by the low-density alluvial deposits and are therefore lower than they would be if the deposits were not present. To compensate for this effect, the initial groundwater basin gravity anomaly was used to calculate an initial estimate of the thickness of the alluvial deposits, and the gravity effect of the groundwater basin was calculated from this initial estimate of the thickness of alluvial deposits at all of the basement gravity stations. A second estimate of the basement gravity field was then made by passing a smooth surface through the corrected basement gravity values; then, a second estimate of the thickness of the alluvial deposits was made. This process was repeated until changes in the calculated thickness of the alluvial deposits were minimal. Note the calculated thickness of the alluvial deposits was constrained by the measured values when they were available.

All calculations were made by using gravity values interpolated to a regular grid having nodes spaced about 820 ft apart, roughly the distance between the gravity stations along the transects. The groundwater basin gravity anomaly was converted to thickness of the alluvial deposits by using an assumed average difference of 300 kg/m³ between the densities of the alluvial deposits that form the groundwater basin and the underlying basement rocks. This density contrast was calibrated by comparing the results of the modeled groundwater basin gravity anomaly to the well depth at locations where the wells fully penetrated the total thickness of the alluvial deposits (*table 1*). Assuming an average density of 2,680 kg/m³ for the pre-Tertiary basement rocks within the study area (median density of 225 samples of granitic, metamorphic, and sedimentary basement rocks from the San Bernardino Mountains and vicinity), the density contrast of 300 kg/m³ between the alluvial deposits and the underlying basement rocks indicates a density of 2,380 kg/m³ for the alluvial deposits. This value is slightly higher than the density (2,300 kg/m³) used by Mabey (1960) to characterize the alluvial deposits in basins in the adjacent Mojave Desert. The higher density suggests lower porosity, perhaps the result of poor sorting of the alluvial deposits that fill the Big Bear Valley groundwater basin.

The resulting density contrast of 300 kg/m³ is reasonable for Quaternary and Tertiary alluvial deposits overlying the pre-Tertiary basement rocks present in the San Bernardino Mountains. This was further tested by examining the basement gravity field for any indications of local anomalies from data at the sites where wells penetrated basement, and the solution was forced to honor those data. Finally, the basement gravity field was compared to the residual magnetic field of the area, which had been mathematically converted to the equivalent of the gravity field, because both fields reflect the distribution of basement rock bodies. This was done in a qualitative way to ensure that no obvious distortions were present in the basement gravity field that were not also present in the converted magnetic field. For example, the shape of the basement gravity high centered over the southeastern edge of the Big Bear Lake surface-water drainage basin was comparable to the shape of the converted magnetic field in the same area, indicating that the gravity field was reasonable in this part of the basin.

Gravity Modeled Thickness of the Groundwater Basin

The gravity model results indicate that the alluvial deposits that form the Big Bear Valley groundwater basin range from less than 500 ft thick on the edges of the groundwater basin and on the eastern end of Big Bear Lake to more than

1,500 ft thick beneath the center of Big Bear Lake and west of Baldwin Lake, with some locations as thick as 2,000 ft (*fig. 6*). Geologic information from 21 water wells (*table 1*; *fig. 6*) was used to constrain the gravity interpretations. Ten of these wells penetrate the entire thickness of the alluvial deposits that form the groundwater basin, and the modeled thicknesses were not allowed to vary in these locations. The modeled thickness of the alluvial deposits and the available well data along east-west and north-south cross sections through the study area are illustrated in *figures 6B* and *6C*. A direct comparison of the modeled and measured thickness of alluvial deposits at 21 wells is included in *table 1*.

Basic uncertainties in the gravity data limit the resolution that can be expected, which even in areas of good measurement coverage is about 50 ft and is less in areas having poor measurement coverage or areas far from either basement outcrop or control points where wells penetrated basement. Additionally, calculations were made on grid cells 820 ft on a side, so the results represent the average thickness of the alluvial deposits at this cell size. Details of the thickness over distances less than a cell-dimension are not resolved. Finally, gravity data, by nature, reflect the average thickness of the causative body and the averaging becomes more pronounced the farther from the source that the observations are made. As a result, if everything else, such as bedrock-surface topographic roughness, were equal, areas where the alluvial deposits are thickest will be subject to higher degrees of averaging and thus will appear smoother than areas where the alluvial deposits are thinner.

The data given in *table 1* indicate that the modeled thickness of the alluvial deposits is reasonable, at least in areas where independent control was available. The modeled thickness of alluvial deposits at places where wells penetrated the entire thickness of these deposits coincide with the measured values to better than 55 ft, but this was expected because the solution was constrained to honor these values. The lack of perfect agreement reflects the spatial averaging discussed above. A better measure of the reliability of the solution can be attained from comparing the modeled thickness values to the total well depths at those wells that did not penetrate the entire thickness of the deposits. The modeled thickness of the alluvial deposits was greater than the total well depth, as it should have been, for all but one of these wells. Only at the Middle School well was the modeled thickness less than the total well depth, and even here the discrepancy was only 18 ft, well within the inherent uncertainty imposed by the basic gravity observations.

Table 1. Locations, available geologic data, and calculated thickness of alluvial deposits from gravity surveys for 21 wells in Big Bear Valley, San Bernardino County, California.

[All data provided by GeoScience Support Services, Inc. **Abbreviations:** ft, feet; °, degrees; ', minutes; –, no data]

Well identification	State well number	Latitude	Longitude	Measured thickness of basin-fill deposits (ft)	Total well depth (ft)	Estimated thickness of basin-fill deposits (ft)	Source
				Wells used to constrain gravity model			
Mallard_Lane	2N1W-24A	34° 14.854'	-116° 55.739'	620	—	647	GeoScience Support Services, Inc. (2004b)
Canvasback	2N/1W-24G	34° 14.772'	-116° 55.956'	450	—	461	GeoScience Support Services, Inc., written commun., 2005
Rathbun	2N/1E-22N	34° 14.461'	-116° 52.288'	980	—	975	GeoScience Support Services, Inc. (2004b)
McAlister	2N/1E-22C	34° 14.905'	-116° 51.921'	690	—	644	GeoScience Support Services, Inc., written commun., 2005
Riffenburgh_BH-1	2N/1E-15F	34° 15.632'	-116° 52.005'	486	—	490	GeoScience Support Services, Inc. (2003b)
Presbyterian_BH-3	2N/1W-24J	34° 14.477'	-116° 55.807'	245	—	232	GeoScience Support Services, Inc. (2003b)
Palomino_Well #8	2N/2E-7P	34° 16.054'	-116° 48.869'	—	609	732	GeoScience Support Services, Inc., written commun., 2005
Test_Hole_E-4	2N/1E-13A	34° 15.852'	-116° 49.422'	—	400	1,396	GeoScience Support Services, Inc., written commun., 2005
Well_3B	2N/1E-12Q2	34° 16.002'	-116° 49.776'	—	820	1,478	GeoScience Support Services, Inc. (2000)
Well 10'	2N/1E-13J	34° 15.437'	-116° 49.483'	634	—	657	GeoScience Support Services, Inc. (2003c)
Maple_Well	2N/1E-24J	34° 14.560'	-116° 49.472'	—	830	972	GeoScience Support Services, Inc., written commun., 2005
La_Crescenta	2N/1E-23Q	34° 14.337'	-116° 50.858'	—	663	1,722	GeoScience Support Services, Inc., written commun., 2005
Sheephorn_TH-5	2N-1E-27A	34° 14.044'	-116° 51.536'	—	529	584	GeoScience Support Services, Inc. (2002)
Greenway_Pk_Well9	2N/1E-14B1	34° 15.782'	-116° 50.629'	—	1,190	1,431	GeoScience Support Services, Inc. (2003d)
Testhole_E2	2N/1E-14C	34° 15.840'	-116° 50.892'	—	498	1,422	GeoScience Support Services, Inc., written commun., 2005
Testhole_E1	2N/1E-11N	34° 15.900'	-116° 51.180'	—	498	954	GeoScience Support Services, Inc., written commun., 2005
Testhole_Northstar	2N/1E-21Q	34° 14.249'	-116° 52.942'	964	—	973	GeoScience Support Services, Inc., written commun., 2005
Testhole_Condor	2N/1E-17Q	34° 15.078'	-116° 53.850'	588	—	584	GeoScience Support Services, Inc., written commun., 2005
Middle_School	2N/1E-20K	34° 14.420'	-116° 54.000'	—	604	586	GeoScience Support Services, Inc., written commun., 2005
Owen	2N/1E-24P	34° 14.196'	-116° 50.074'	—	1,070	1,155	GeoScience Support Services, Inc. (1999)
Metcalf	2N/1W-24P	34° 14.334'	-116° 56.166'	175	—	170	GeoScience Support Services, Inc., written commun., 2005
				Wells not used to constrain gravity model			
Pennsylvania #1	2N/1E-19J2	34° 14.449'	-116° 54.580'	652	—	508	Department of Water Resources, written commun., 2005
Knickerbocker	2N/1E-19H	34° 14.652'	-116° 54.554'	—	825	1,000	Department of Water Resources, written commun., 2005
Oak well #9	2N/1E-20R1	34° 14.388'	-116° 53.724'	838	—	755	Department of Water Resources, written commun., 2005
Lassen #4	2N/1E-27J	34° 13.541'	-116° 51.490'	220	—	167	Department of Water Resources, written commun., 2005
Division Well #1	2N/1E-15C2	34° 15.737'	-116° 52.030'	—	423	499	Department of Water Resources, written commun., 2005
Division Well #2	2N/1E-15C1	34° 15.786'	-116° 51.997'	—	497	499	Department of Water Resources, written commun., 2005
Division Well #3	2N/1E-15C	34° 15.739'	-116° 52.068'	473	—	499	Department of Water Resources, written commun., 2005
Division Well #4	2N/1E-15C	34° 15.912'	-116° 52.096'	—	475	499	Department of Water Resources, written commun., 2005
Division Well #7	2N/1E-15C	34° 15.738'	-116° 51.991'	399	—	322	Department of Water Resources, written commun., 2005
Bow Canyon	2N/1E-26E1	34° 13.751'	-116° 51.146'	516	—	650	Department of Water Resources, written commun., 2005
Magnolia	2N/1E-24G	34° 14.607'	-116° 49.855'	—	1,140	985	GeoScience Support Services, Inc., written commun., 2005
Moonridge	2N/1E-21J	34° 14.557'	-116° 52.708'	—	1,135	985	GeoScience Support Services, Inc., written commun., 2005

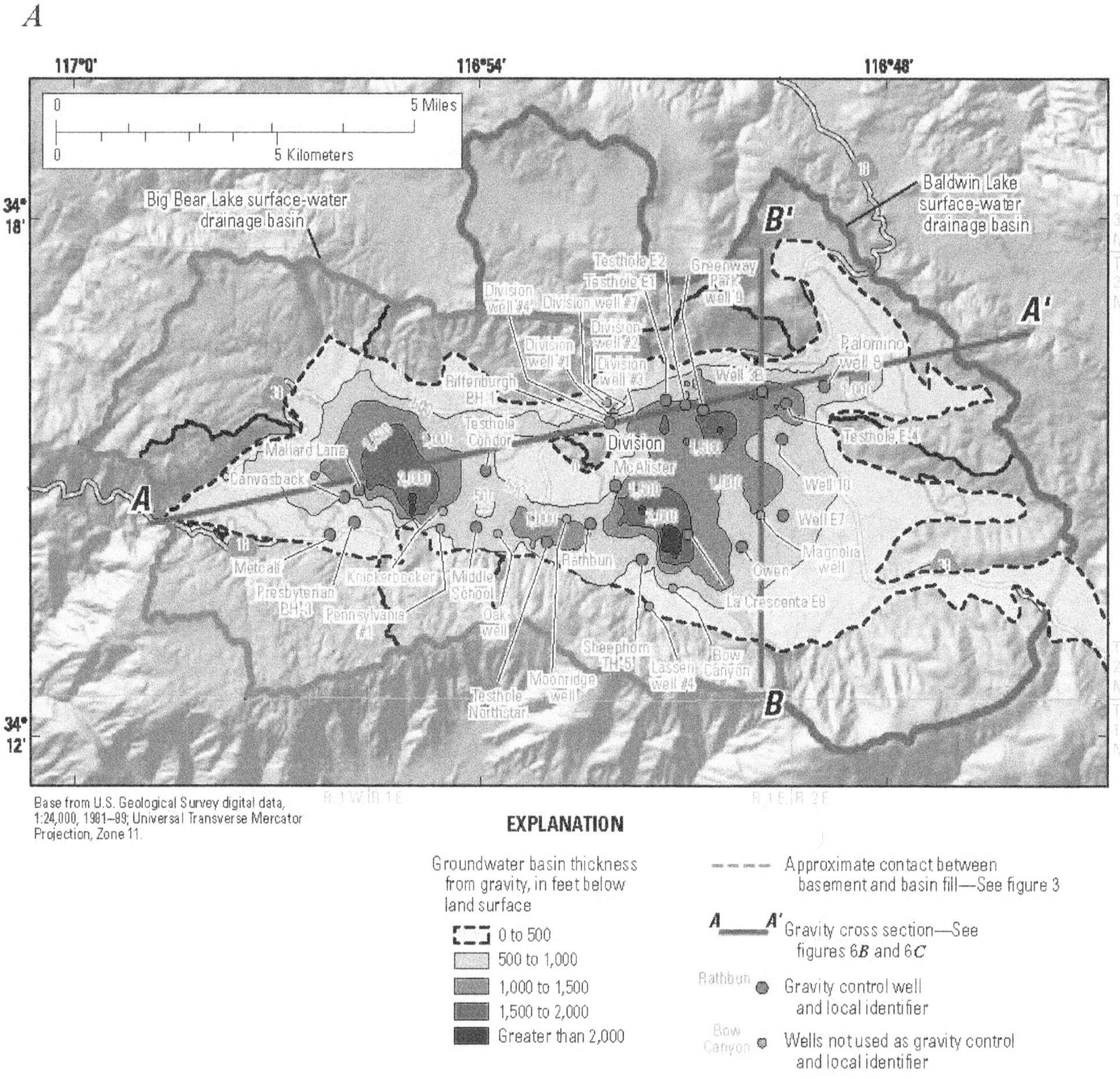

A

Base from U.S. Geological Survey digital data,
1:24,000, 1981–89; Universal Transverse Mercator
Projection, Zone 11.

EXPLANATION

Groundwater basin thickness
from gravity, in feet below
land surface

⌐⌐⌐	0 to 500
▢	500 to 1,000
▨	1,000 to 1,500
▨	1,500 to 2,000
▨	Greater than 2,000

– – – –   Approximate contact between
basement and basin fill—See figure 3

A ——— *A'* Gravity cross section—See
figures 6*B* and 6*C*

Rathbun ● Gravity control well
and local identifier

Bow
Canyon ● Wells not used as gravity control
and local identifier

Figure 6.   Thickness of the alluvial deposits based on gravity data (*A*) in the Big Bear Valley groundwater basin, (*B*) along section A-A′, and (*C*) along section B-B′, and (*D*) altitude of the top of the basement complex calculated from gravity measurements in the Big Bear Valley, San Bernardino, California. Click on figure 6D to see and control animation showing the altitude of the top of the basement complex.

Structure of the Groundwater Basin

The gravity data indicate that the alluvial deposits that fill the groundwater basin are thickest beneath the center of Big Bear Lake and the area west of Baldwin Lakes (*fig. 6*). In these areas, the alluvial deposits have a maximum thickness of more than 2,000 ft. The modeled thickness of the alluvial deposits between Big Bear and Baldwin Lakes could be less than that shown on *figure 6* because very low-density Quaternary alluvial deposits mantle the denser older Tertiary

sedimentary deposits that are present at depth in this part of the basin. These very low-density deposits were not specifically taken into account when modeling the thickness of the alluvial deposits. Not considering these low-density deposits explicitly would result in the modeled values being thicker than they actually are. Only detailed independent information on the specific thickness of these Quaternary alluvial deposits, such as borehole data, would permit a more accurate solution in this area.

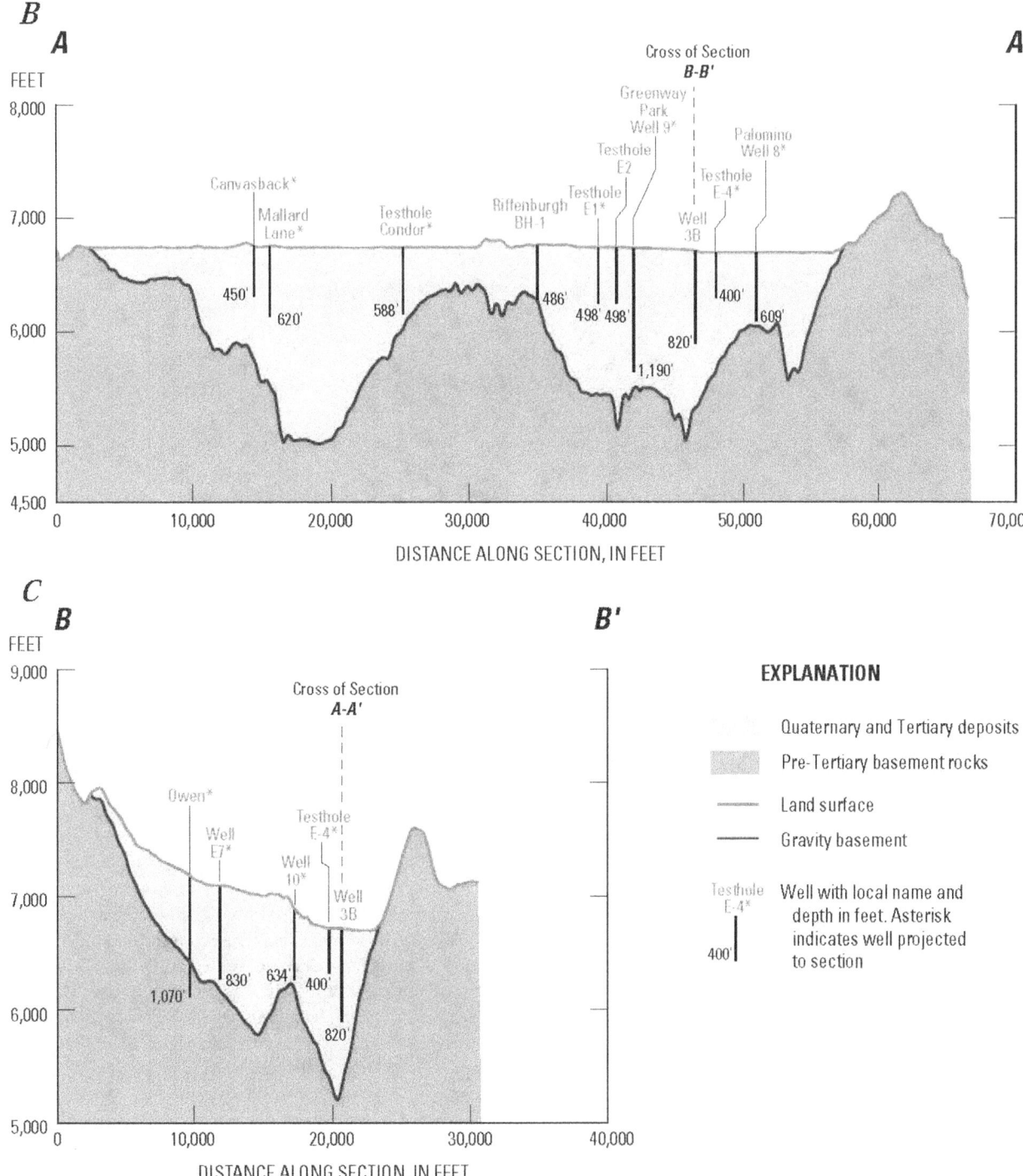

Figure 6.   Continued

D

Click on the figure to see and control animation.

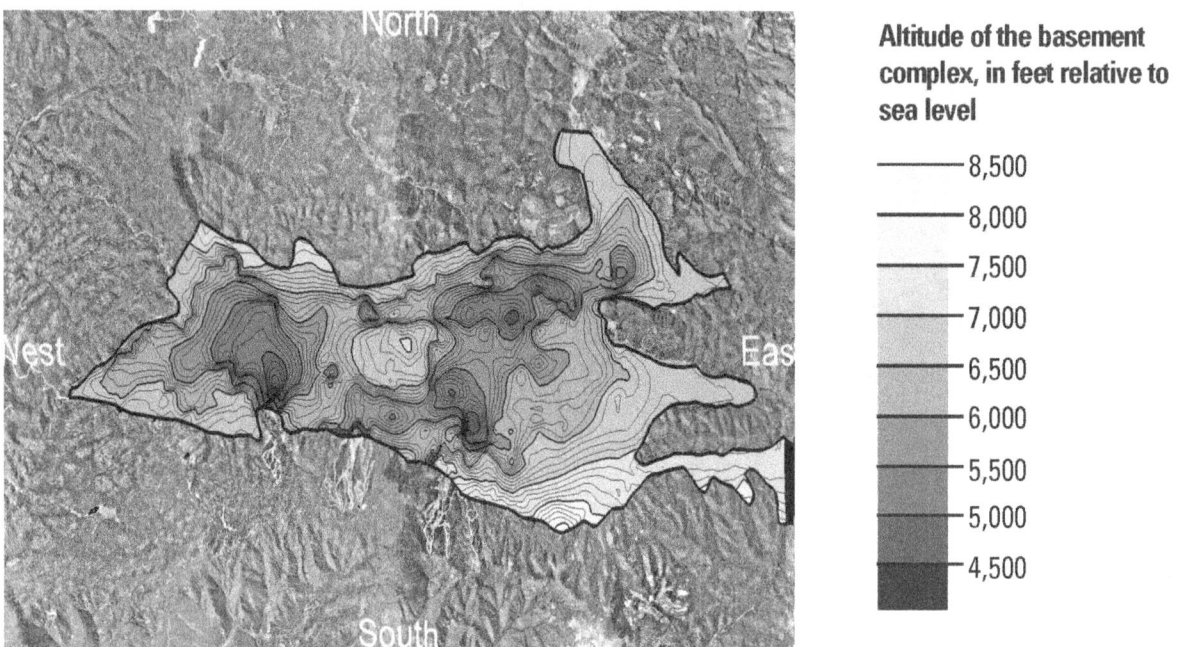

Altitude of the basement complex, in feet relative to sea level

——— 8,500
——— 8,000
—— 7,500
— 7,000
— 6,500
— 6,000
— 5,500
— 5,000
— 4,500

Figure 6. Continued

Gravity data do not provide definitive evidence of faults that might affect groundwater flow; however, large changes in gravity gradients can be used to infer the presence of a fault. In the Division subbasin, the gravity data are interpreted to indicate a large change in the thickness of alluvial deposits along the western edge of the subbasin (*fig. 6A*). This change in thickness suggests the presence of a northwest-southeast trending fault, although a fault has not been mapped in this area.

To help visualize the basin geometry in the Big Valley groundwater basin, an animation of the altitude of the top of the basement-complex was prepared (*fig. 6D*). The altitude of the top of the basement complex was calculated by subtracting the modeled thickness of the valley-fill deposits at each gravity grid from the average land-surface altitude at that grid. The animation allows the viewer to fly over the ridges and valleys of the shaded-relief altitude of the top of the basement complex.

Relation of Calculated Groundwater-Basin Thickness to Groundwater Availability

As stated previously, the alluvial deposits that form the groundwater basin consist of Tertiary sedimentary deposits

and Quaternary alluvial deposits. The Tertiary sedimentary deposits consist primarily of consolidated to partly consolidated alluvial-fan deposits and probably yield only small quantities of water to wells. The Quaternary alluvial deposits consist of interbedded layers of water-bearing sand and gravel and non-water-bearing silt and clay. The gravity method used for this study does not differentiate between water-bearing and non-water-bearing deposits; therefore, the calculated combined thickness of these deposits cannot be used independently to estimate the groundwater availability in the groundwater basin. Only detailed independent information on the specific thickness and water-bearing properties of these different deposits, which could be provided by borehole data, would permit an accurate estimate of groundwater availability. The thickness map prepared for this study could be used to help identify the location of potential boreholes to investigate the water-bearing properties of the groundwater basin. For example, areas on the map where the alluvial deposits are identified as having a substantial thickness, and where there is no existing geologic information, could be good locations to drill exploratory boreholes.

Interferometric Synthetic Aperture Radar (InSAR)

By Michelle Sneed, and Justin Brandt

Interferometric Synthetic Aperture Radar (InSAR) was used in this study to measure pumping-induced land subsidence and locate structures, such as faults, that can affect groundwater movement. A description of the mechanics of pumping-induced land subsidence is included to provide background information.

Mechanics of Pumping-Induced Land Subsidence

Land subsidence is known to occur in basins containing aquifer systems that at least in part consist of fine-grained sediments and have undergone extensive groundwater development. The pore structure of a sedimentary aquifer system is supported by a combination of the granular skeleton of the aquifer system and the pore-fluid pressure of the groundwater that fills the intergranular pore space (Meinzer, 1928). Constant total stress on the aquifer system is equivalent to a constant total weight of the overlying sediments and pure fluid—the overburden. When an aquifer is under constant total stress and groundwater is withdrawn in quantities that result in reduced pore-fluid pressures and water-level declines, reduced pore-fluid pressure support increases the intergranular stress, or effective stress, on the skeleton. A change in effective stress deforms the skeleton—an increase causes some degree of skeletal compression, and a decrease causes some degree of expansion. The vertical component of skeletal compression sometimes results in irreversible compaction of the aquifer system and land subsidence. An aquifer system that primarily consists of fine-grained sediments, such as silt and clay, is much more compressible than one that primarily consists of coarse-grained sediments, such as sand and gravel.

Aquifer-system deformation is elastic (recoverable) if the effective stress imposed on the skeleton is smaller than any previous effective stress (Leake and Prudic, 1991). The largest historical effective stress imposed on an aquifer system—sometimes as a result of the lowest groundwater level—is called the "preconsolidation stress." If a stress imposed on the skeleton is greater than the preconsolidation stress, the pore structure of the granular matrix of the fine-grained sediments is rearranged; this new configuration results in a reduction of pore volume and, thus, inelastic (largely irreversible) compaction of the aquifer system. Furthermore, the compressibility of the fine-grained sediments under stresses greater than the preconsolidation stress is 20 to more than 100 times greater than under stresses less than the preconsolidation stress (Riley, 1998). Inelastic compaction of coarse-grained sediment is negligible.

A significant part of the total compaction of an aquifer-system skeleton that contains an appreciable thickness of fine-grained sediments can be residual compaction, or delayed compaction that occurs in thick fine-grained interbeds and confining layers while heads equilibrate with heads in the adjacent aquifers (Terzaghi, 1925). Depending on the thickness and the vertical hydraulic diffusivity (transmissivity divided by storage) of a confining layer, pressure equilibration—and thus compaction—can lag behind pressure, or head, changes in the adjacent aquifers; this lag can be on the order of centuries. For a more complete description of aquifer-system compaction, see Poland (1984), and for a review and selected case studies of land subsidence caused by aquifer-system compaction in the United States, see Galloway and others (1999).

InSAR Methodology

InSAR is a satellite-based remote-sensing technique that measures vertical changes of land-surface elevation with a resolution of less than 0.5 in. under good conditions. The InSAR technique uses two Synthetic Aperture Radar (SAR) images of the same area acquired at different times and "interferes" (differences) them, resulting in maps called interferograms that show line-of-sight ground-surface displacement (range change) between the two imaging times. SAR imagery is produced by reflecting radar signals off a target area and measuring the two-way travel time from and to the satellite. Generating an interferogram produces two components: amplitude and phase. The amplitude component is the measure of the radar signal intensity returned to the satellite and shows buildings, roads, mountains, and other reflective features; the phase component is proportional to range change and shows the coherent displacements imaged by the radar. If the ground has moved away from (subsidence) or toward (uplift) the satellite between the times of the two acquisitions, a slightly different portion of the wavelength is reflected back to the satellite resulting in a measurable phase shift that is proportional to range change. The map of phase shifts, or interferogram, is depicted with a repeating color scale that shows relative range change between the first and the second acquisitions; in this report, one complete color cycle (fringe) represents 1.1 in. of range change. The indicated range change is about 90-95 percent of true vertical ground motion, depending on the satellite look angle and location of the target area. The direction of change—subsidence or uplift—is indicated by the color progression of the fringe(s) toward the center of a deforming feature. For interferograms in this report, the color-fringe progression of blue-green-yellow-orange-red-purple indicates subsidence; the opposite progression indicates uplift.

InSAR signal quality can produce "noise" that is significantly greater than the less-than-centimeter resolution so that it masks deformation because signal quality, in part, depends on satellite position, atmospheric effects, ground cover, land-use practices, and temporal separation of the interferogram. Strict orbital control is required to precisely control the look angle and position of the satellite. Applying the InSAR technique successfully is contingent on looking at the same point on the ground from the same position in space, such that the horizontal distance between each satellite pass, or perpendicular baseline, is minimized. Perpendicular baselines greater than about 650 ft produce excessive topographic effects (parallax) that can mask real signal. Phase shifts can be caused by variable atmospheric mass that is associated with different elevations. A digital elevation model (DEM) is used in the interferogram generation process to reduce the effects caused by elevation differences (and to georeference the image, as well). Phase shifts also can be caused by laterally variable atmospheric conditions, such as clouds or fog, because the non-uniform distribution of water vapor differentially slows the radar signal over an image. Atmospheric artifacts can be identified by using several independent interferograms, which are defined as interferograms that do not share a common SAR image. When apparent ground motion is detected in only one interferogram, or in a set of interferograms sharing a common SAR image, the apparent motion likely is due to atmospheric phase delay and can be discounted. Many interferograms should be inspected for a study. If a signal appears repeatedly in independent interferograms, it is believable. If it appears only in dependent interferograms, it is questionable at best, and it would not be interpreted as deformation.

The type and density of ground cover also can affect interferogram quality significantly. Densely forested areas, such as those in the Big Bear area, are prone to poor signal quality because the C-band wavelength (2.2 in.) cannot effectively penetrate thick vegetation and is either absorbed or reflected back to the satellite from varying depths within the canopy, resulting in an incoherent signal (shown as randomized colors on an interferogram). Sparsely vegetated areas and urban centers, however, generally have high signal quality because bare ground, roads, and buildings have high reflectivities and are relatively uniform during at least some range of InSAR timescales. Certain land-use practices, such as farming, also cause incoherent signal return. The tilling and plowing of farm fields causes large and nonuniform ground-surface change that cannot be resolved with InSAR. Signal quality is adversely affected by larger temporal separations, also, because there is more opportunity for nonuniform change in both urban and non-urban areas. Many of these error sources

were minimized by producing interferograms having perpendicular baselines less than 650 ft and by examining several independent interferograms for the area of interest in the Big Bear area, which is fairly flat and contains several urban centers.

For this study, SAR data from the European Space Agency's (ESA) ERS-1, ERS-2, and ENVISAT satellites were used to map and measure range change. The singular-mission, twin satellites, ERS-1 and ERS-2, were launched in 1991 and 1995, respectively; ERS-1 was turned off in 1999, and ERS-2 has not been routinely suitable for interferometric applications since late 2000. The multi-mission ENVISAT platform was launched in 2002 and currently is the only ESA-owned fully functional SAR satellite. The ESA satellites provided data for 1992–2000, and the ENVISAT satellite provided data for 2003 to 2008. The three satellites are side-looking, orbit the Earth at an altitude of approximately 500 mi, and have 35-day repeat cycles. Seventeen ERS-1 and ERS-2 SAR images were used to produce eleven interferograms for 1992–2000 (for example, see *fig. 7A*), and nine ENVISAT SAR images were used to produce five interferograms for 2004–2005 (for example, see *fig. 8A*). The images have temporal baselines ranging from 35 to 980 days between December 25, 1992, and May 30, 2005.

InSAR Calculations of Land-Surface Deformation

Inspection of the 16 interferograms (*table 2A*) developed for this study by using the techniques and criteria discussed above indicated that land-surface deformation has occurred in the Big Bear Valley groundwater basin. Three general deforming areas were identified: (1) the area between Big Bear Lake and Baldwin Lake, (2) the area near the city of Big Bear Lake, and (3) the area near Sugarloaf (for example, see *figs. 7* and *8; table 2A*). Available water-level data provided by CSD and DWP from wells in these areas (*fig. 9; table 2B*) were used to determine if there was a relation between water-level change and land-surface deformation. Relating concurrent water-level changes and deformation in these areas is difficult because of two factors. Many of the wells with sufficient water-level data to compare with InSAR results have fairly long or multiple-screened intervals, or both, making it impossible to relate depth-integrated water-level data to a potential effect on depth-specific sediments. Additionally, thick clay deposits (in excess of 50-ft thick), such as those found in the Big Bear Valley groundwater basin, generally have complicated and lagged responses to pore-pressure changes, as discussed in the section describing the mechanics of pumping-induced land subsidence.

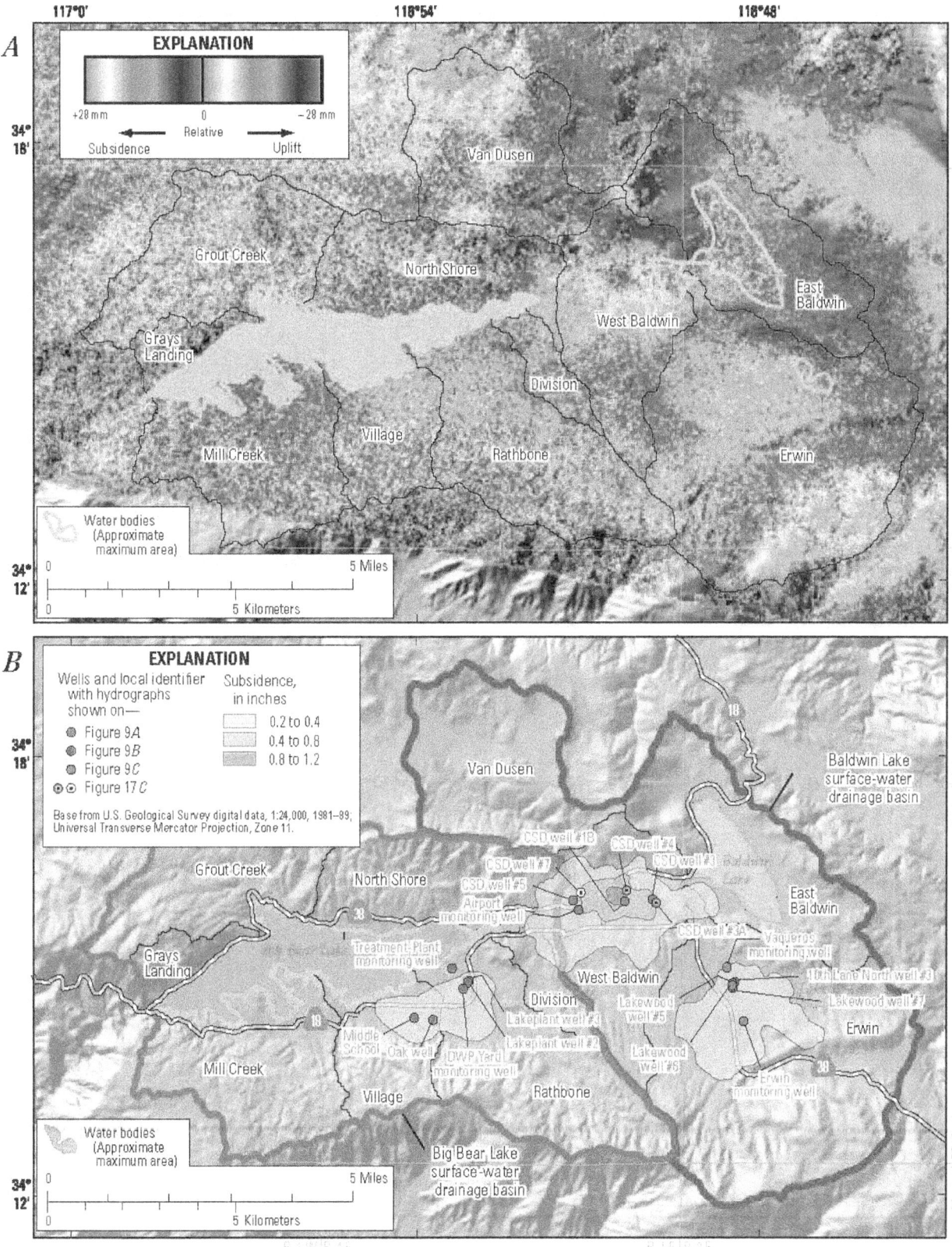

Figure 7. (*A*) Interferogram showing vertical changes in land surface in Big Bear Valley, San Bernardino County, California, between September 25, 1995 and July 21, 1997, and (*B*) corresponding calculation of subsidence for that period. White areas on interferogram indicate no data.

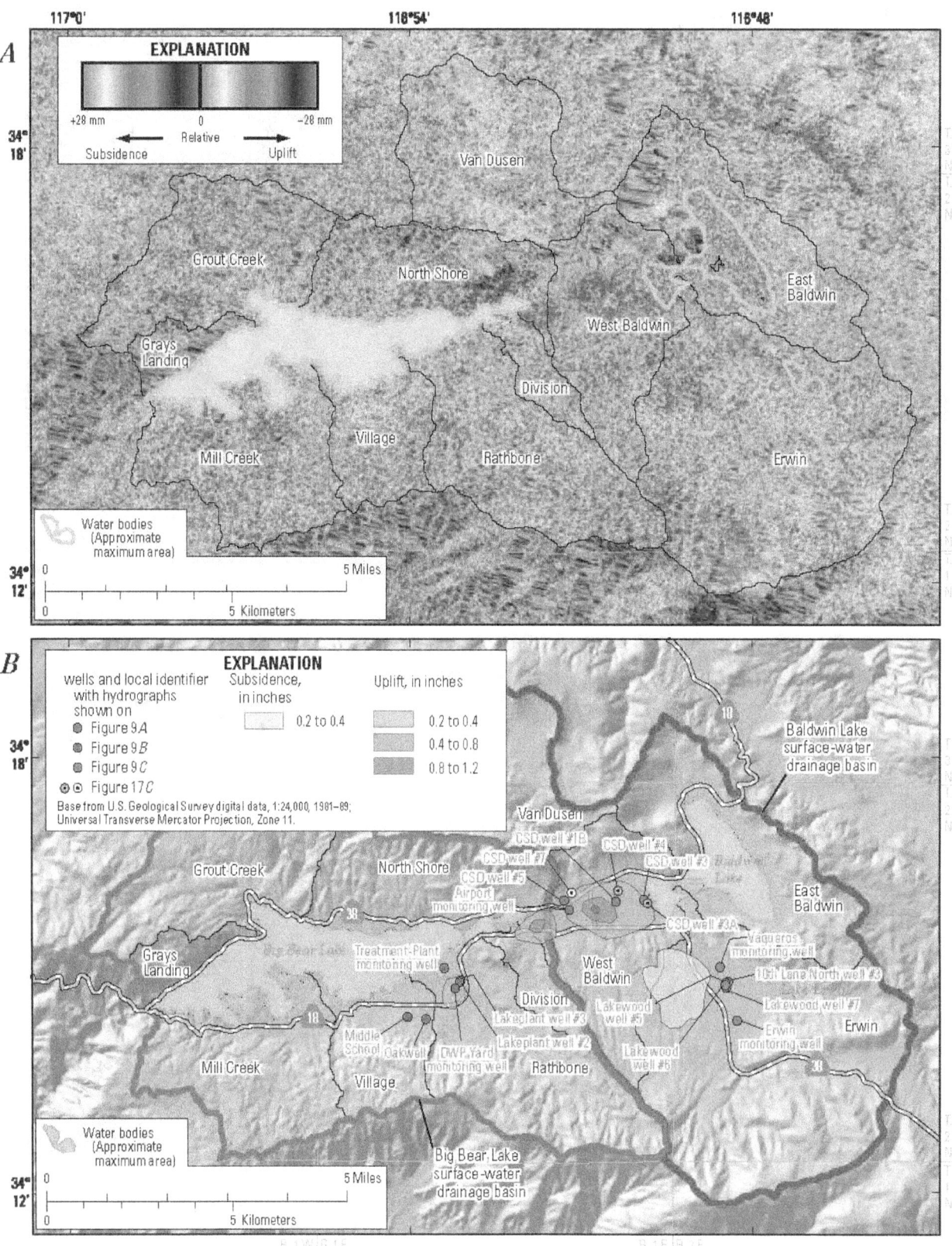

Figure 8. (*A*) Interferogram showing vertical changes in land surface in Big Bear Valley, San Bernardino County, California, between August 23, 2004 and May 30, 2005, and (*B*) corresponding calculation of subsidence for that period.

Table 2. (**A**) Acquisition dates of 16 interferograms and corresponding deformation for three locations in Big Bear Valley, and (**B**) wells used for water-level analyses in Big Bear Valley, San Bernardino County, California.

[**Abbreviations**: mm/dd/yyyy, month/day/year; N/A, not available; ND, no deformation detected; +, uplift; –, subsidence]

(**A**)

	Interferogram		Deformation, in inches		
Interferogram identification	Acquistion date 1 (mm/dd/yyyy)	Acquistion date 2 (mm/dd/yyyy)	Between Big Bear and Baldwin Lakes	Near city of Big Bear Lake	Near Sugarloaf
1	12/25/1992	08/27/1993	+0.4	ND	ND
2	06/18/1993	06/11/1995	–0.4	–0.4	+0.2
3	09/25/1995	03/18/1996	+0.8	ND	ND
4	09/25/1995	07/21/1997	–1.0	–0.4	–0.4
5	09/25/1995	06/01/1998	–0.6	ND	ND
6	03/18/1996	06/16/1997	–0.4	–0.2	ND
7	06/16/1997	06/01/1998	+0.4	+0.4	ND
8	11/23/1998	08/30/1999	–0.6	–0.2	–0.4
9	07/26/1999	03/27/2000	–1.0	–0.4	–0.6
10	11/08/1999	11/27/2000	–0.4	ND	ND
11	06/05/2000	11/27/2000	–0.2	ND	ND
12	07/16/2004	09/24/2004	–1.0	ND	ND
13	04/05/2004	04/24/2005	–1.2	ND	–0.2
14	05/10/2004	03/21/2005	–1.2	ND	ND
15	08/23/2004	05/30/2005	+1.2	+0.2	–0.2
16	02/14/2005	03/21/2005	+0.8	+0.2	+0.4

(**B**)

State well number	Site identification	Easting (meters)	Northing (meters)
3610008-011	CSD well #5	513,356.30	3,791,781.30
3610008-009	CSD well #4	514,805.90	3,792,028.50
3610008-005	CSD well #1B	514,751.10	3,791,751.30
3610008-007	CSD well #3	515,505.90	3,791,807.90
3610008-008	CSD well #3A	515,598.10	3,791,719.40
N/A	Airport monitoring well	513,490.20	3,791,538.70
N/A	Treatment plant monitoring well	510,140.60	3,790,059.80
3610044-035	Middle school	509,129.00	3,788,810.20
3610044-038	Oak well	509,631.90	3,788,750.80
3610044-011	Lakeplant well #2	510,592.20	3,789,742.00
3610044-012	Lakeplant well #3	510,564.60	3,789,702.10
N/A	Vaqueros monitoring well	517,535.50	3,790,086.20
3610061-010	Lakewood well #7	517,713.20	3,789,564.30
3610061-001	10th Lane north well #3	517,753.50	3,789,706.30
3610061-006	Lakewood well #5	517,702.90	3,789,678.50
3610061-007	Lakewood well #6	517,684.60	3,789,631.90
N/A	Erwin monitoring well	517,997.60	3,788,718.90
N/A	DWP yard monitoring well	510,444.20	3,789,546.70

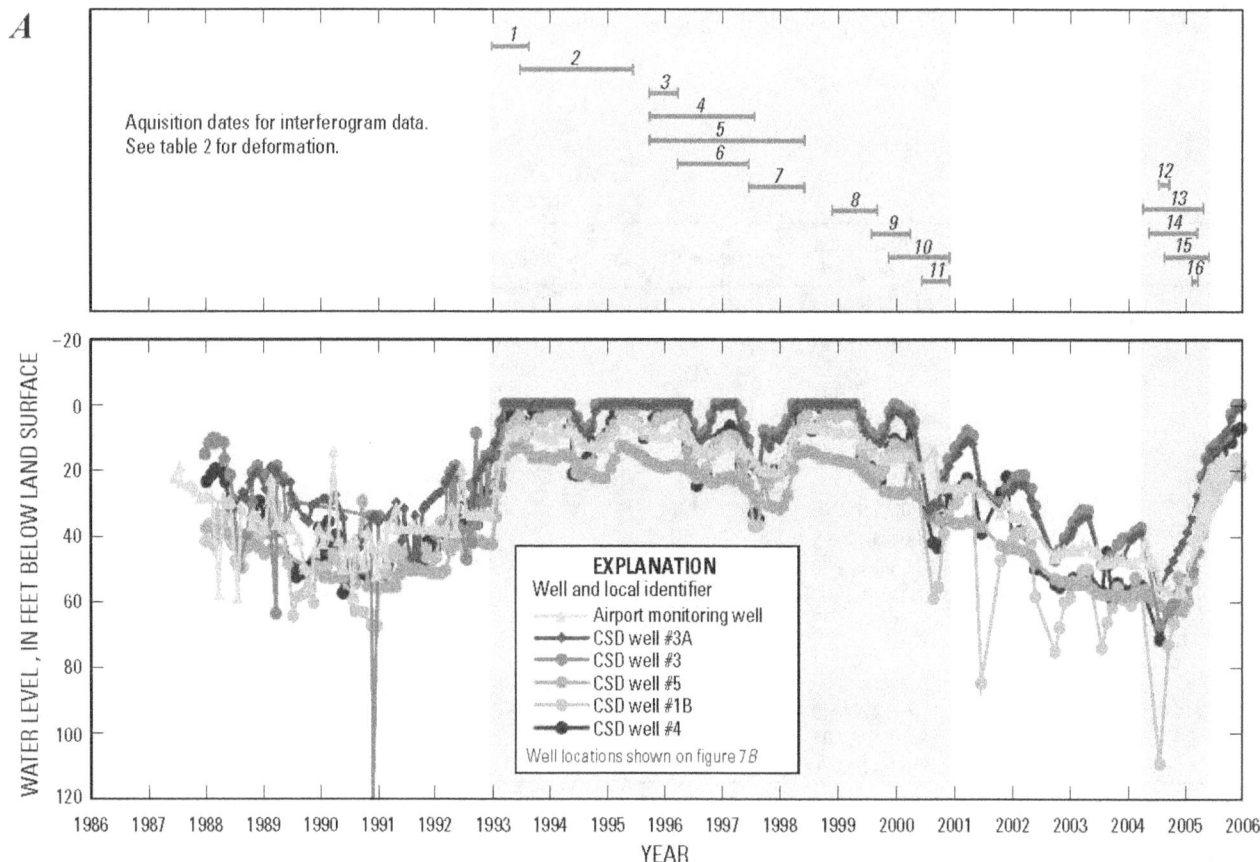

Figure 9. Hydrographs showing water levels provided by Big Bear Community Services District (CSD) and Big Bear Lake Department of Water and Power (DWP) for selected wells in areas where interferograms show deformation in the area (*A*) between Big Bear and Baldwin Lakes, (*B*) near the city of Big Bear Lake, and (*C*) near Sugarloaf in Big Bear Valley, San Bernardino County, California.

In the area between Big Bear and Baldwin Lakes, eleven interferograms indicate land-surface subsidence, and five interferograms indicate uplift for various time periods between December 25, 1992, and May 30, 2005 (for example, see *figs. 7* and *8*; *table 2A*). About half of the interferograms show maximum deformation near the western extent of Baldwin Lake; whereas, other interferograms show maximum deformation near the eastern extent of Big Bear City Airport. Cross sections constructed using lithologic and geophysical logs of boreholes in the area between Big Bear and Baldwin Lakes indicate that the groundwater basin is composed of sediments with a substantial fraction of silt and clay (Thomas Harder, GeoScience Support Services, Inc., written commun., 2005), implying that subsidence in this area has the potential to be inelastic, or permanent. Water levels collected by CSD and DWP during 1992–2000 did not drop below the lows of 1990 and 1991 (*fig. 9A*) and, therefore, did not exceed the preconsolidation stress during this 8-year period. These data indicate either the subsidence that occurred during this time

was elastic and responding to seasonal water-level fluctuations, or the subsidence was residual and mostly inelastic, responding to the water-level lows of 1990 and 1991 or earlier unknown lows (*figs. 7* and *9A*; *table 2A*). The interferograms for 2004–05 show both subsidence and uplift during a period of net water-level rise (*fig. 9A*). Three of the interferograms showed 1.0 or more inches of subsidence during periods when water-levels rose as much as 20 ft. Subsidence that occurred when water levels were rising indicates that the subsidence was residual and occurred in response to the water-level lows between 2000 and mid-2004, when levels were the lowest for the period of record, 1986–2006 (*fig. 9A*). If these water-level lows exceeded the preconsolidation stress that could have been set before 1986, or defined the preconsolidation stress with these new low levels, then the subsidence is largely inelastic. In the other two interferograms, 0.8 or more inches of uplift were detected concurrent with both large (50 ft) and small (5 ft) water-level increases (for example, see *fig. 8*).

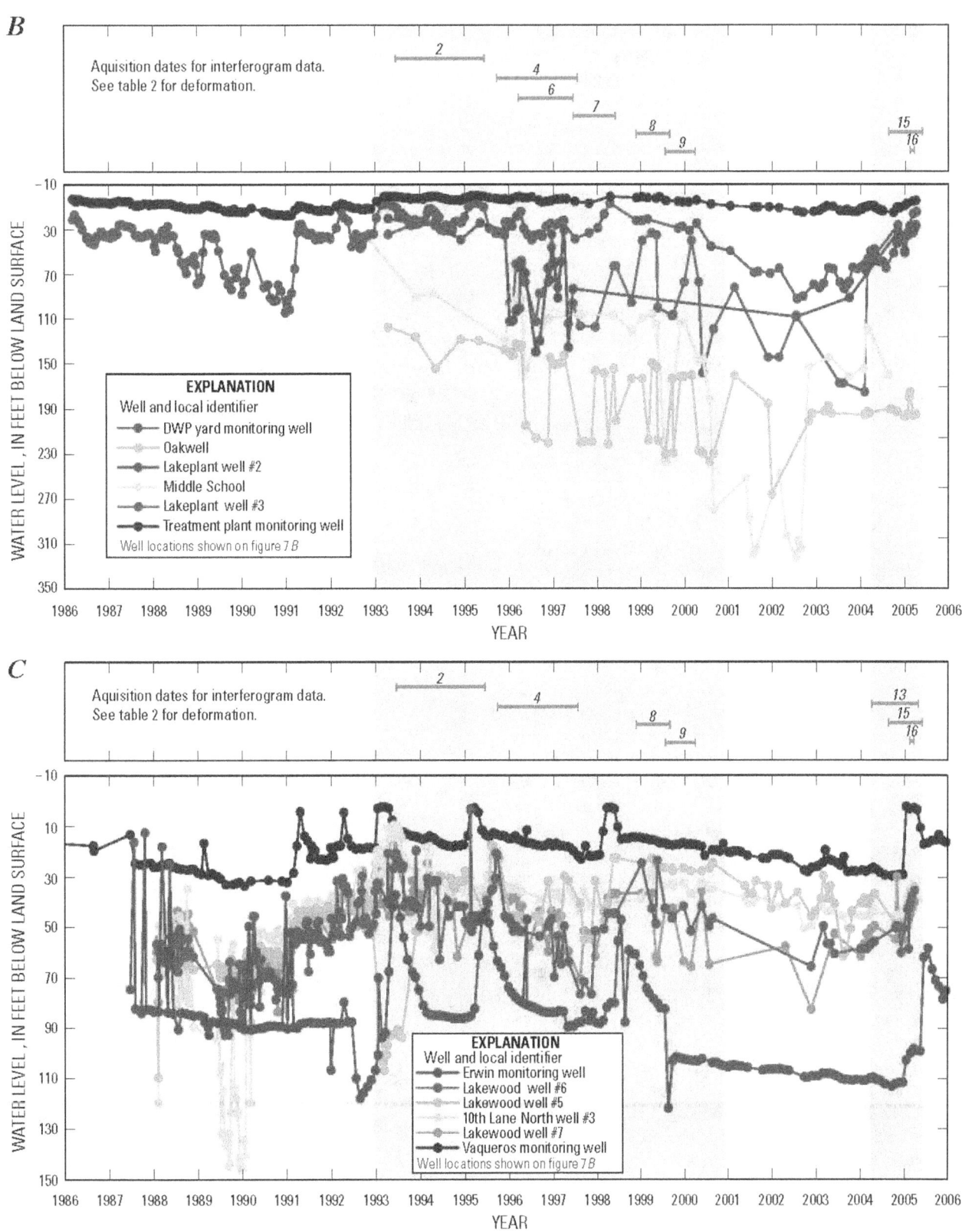

Figure 9. Continued

The changes in land-surface deformation observed in the 2004–05 interferograms can be explained by variable thicknesses of silt and clay layers. When water levels started rising in October 2004, thin layers of silt and clay expanded, but not enough to mask the magnitude of continuing residual compaction occurring in thick layers of silt and clay responding to previous water-level lows. As the water levels continued to rise in 2005, the expansion of the thin silt and clay layers and the coarser grained sand and gravel layers exceeded the residual compaction in the thick silt and clay layers. This can explain why compaction was detected in earlier periods and expansion was detected in later periods in the five temporally overlapping 2004–05 interferograms. Compaction of thick silt and clay layers concomitant with expansion in other portions of an aquifer system has been deduced for other basins in California and Nevada (Sneed and Galloway, 2000; Pavelko, 2004).

In the area near the city of Big Bear Lake, five interferograms showed subsidence and three interferograms showed uplift for various time periods between June 18, 1993, and May 30, 2005 (for example, see *figs. 7* and *8*; *table 2A*). The location of maximum deformation is elongate west-east and centered along Highway 18 (*fig. 7*). Cross sections constructed from lithologic and geophysical logs of boreholes near the city of Big Bear Lake indicate that the underlying 400–500-ft-thick sediments contain a substantial fraction of silt and clay (Thomas Harder, GeoScience Support Services, Inc., written commun., 2005). Water-level data provided by DWP near the city of Big Bear Lake are available beginning with the year 1986; however, most of the measurements began in 1993 (*fig. 9B*). Most water levels show seasonal and other shorter-term water-level changes for the period of record. On longer time scales, however, water levels in some wells generally decreased between 1993 and 2005 (for example, Oak well in *figure 9B*); whereas, water levels in other wells rose in the early 1990s, changed little through late 2002, then rose from late 2002 through 2005 (for example, DWP yard monitoring well in *figure 9B*). It is unclear if the small amount of deformation in this area is elastic or inelastic because the water-level record is incomplete before 1993.

Subsidence in the Sugarloaf area was identified in five interferograms, and uplift was identified in two interferograms for various periods between June 18, 1993, and May 30, 2005 (for example see *figs. 7* and *8*; *table 2A*). Cross sections constructed from lithologic and geophysical logs of boreholes in the Sugarloaf area indicate the predominance of coarse-grained sand and gravel deposits in this part of the groundwater basin (Thomas Harder, GeoScience Support Services, written commun., 2005). The predominance of coarse-grained deposits indicates that deformation in this area is elastic only. Water levels provided by CSD in the Sugarloaf area show seasonal and other shorter-term fluctuations but are relatively stable over longer periods. In general, the water-level changes

follow the pattern described for the other areas (*fig. 9*). Water levels in many wells in this area were historically low in the late 1980s and early 1990s, but water levels since that time generally have not declined to levels below the historical lows. The geologic and water-level data suggest that stresses in the Sugarloaf area are in the elastic range, and that the small amounts of deformation in the area are recoverable.

InSAR Inferred Structure

Measurements of land-surface subsidence can be used to infer the location of buried faults not readily evident on the surface (Galloway and others, 1999). In alluvial basins, faults are commonly barriers to groundwater flow (Galloway and others, 1999); therefore, water-level changes and related land-surface deformation are greater on the side of the fault where pumping occurs. Some of the interferograms show a northwest-southeast trending linear feature between Big Bear and Baldwin Lakes, indicating the presence of a fault or an abrupt change in lithology, although without further investigation it is not possible to confirm the existence of a fault.

Groundwater Recharge in the Big Bear Valley

The quantity, distribution, and source of groundwater recharge in the Big Bear Valley were estimated by using two numerical modeling approaches: (1) a regional-scale Basin Characterization Model (BCM; Flint and others, 2004) and (2) a drainage-basin scale precipitation-runoff model, INFILv3 (Hevesi and others, 2003; U.S. Geological Survey, 2008). The BCM uses average monthly climate data and originally was developed to estimate recharge and runoff for multiple basins throughout the desert southwest. Model results are useful for bounding water-balance results of more detailed models, evaluating long-term climate conditions, illustrating the mechanisms responsible for recharge in a basin, and comparing the locations and volumes of recharge and runoff in different basins on a regional scale. The INFILv3 model uses daily climate data and a water-balance model of the root zone with a primarily deterministic representation of the processes controlling recharge. The water balance includes stream-flow routing and simulated evapotranspiration. Evapotranspiration is simulated as a function of the vertical distribution of available water in a multi-layered root-zone.

For this study, the INFILv3 model was modified to include a shallow groundwater zone (SGWZ) underlying the root zone. The SGWZ was used to model lateral groundwater seepage and changes in water content of the soil and bedrock below the root. Net infiltration is defined as inflow to the SGWZ, and recharge is defined as vertical percolation through the bottom of the SGWZ. A second modification of the original INFILv3 model was the addition of a simplified water-balance model to simulate changes in the levels and volumes of Big Bear and Baldwin Lakes. The simplified water-balance model for the lakes (referred to as the LAKE model) consists of a post-processing routine that uses the daily INFILv3 results to simulate lake levels and volumes. The simulated daily lake levels and volumes were compared to available records for the purpose of calibrating the combined INFILv3-LAKE model.

The BCM and INFILv3 models produce spatially- and temporally-distributed results that represent all components of the water balance, which improve understanding of the primary mechanisms responsible for recharge in a basin. Used together, the BCM and INFILv3 models are complementary and provide a reasonable check on simulated recharge values because of the differences in the underlying assumptions of both models. In this application, the models provide a range of recharge values that bound the likely recharge and reflect the uncertainties associated with a complex system with limited data available for calibration.

Previous Estimates of Recharge

Previous studies have estimated maximum perennial yield, and in some cases recharge, for the Big Bear Valley by using various approaches. Maximum perennial yield is defined by Todd (1980) to be the "maximum quantity of groundwater available perennially if all possible methods and sources are developed for recharging the basin." Maximum perennial yield is a subset of total groundwater recharge because not all water recharged can be developed for beneficial use. Although not exactly comparable to recharge, perennial-yield estimates provide a general estimate of water availability.

Big Bear Lake Surface-Water Drainage Basin

GeoScience Support Services, Inc. (2001) estimated that the maximum perennial yield for the Big Bear Lake surface-water drainage basin ranged from 2,820 to 2,970 acre-ft/yr using a combination of Darcian flow calculations (Roscoe-Moss, 1990) and water-balance techniques (Roscoe-Moss, 1990). The Rathbone subbasin was determined to have the largest maximum perennial yield, ranging from 1,100 to 1,200 acre-ft/yr.

GeoScience Support Services, Inc. (2003) estimated recharge and runoff for the North Shore and Grout Creek subbasins of the Big Bear Lake surface-water drainage basin

(*fig. 1*) by using the Hydrologic Simulation Program–FORTRAN (HSPF; Bicknell and others, 1997), a distributed parameter watershed model. This model requires site-specific topographic data, plus inputs for upper boundary conditions (precipitation, air temperature) and values for hydraulic parameters that control infiltration and runoff. The HSPF model typically is used to simulate surface-water flow. The model has a deep percolation calibration parameter that controls the amount of simulated surface water that percolates and becomes recharge. This parameter usually is adjusted during the model-calibration process to improve the match between measured and simulated streamflow. The HSPF model developed for the North Shore and Grout Creek subbasins incorporated a range of values for hydraulic parameters that were used in published models from across the nation instead of calibrating the parameters by using data collected in Big Bear Valley (GeoScience Support Services, Inc., 2003). Model results indicated that the North Shore subbasin had an average recharge of about 290 acre-ft/yr, and the Grout Creek subbasin had an average recharge of about 550 acre-ft/yr.

GeoScience Support Services, Inc. (2001) refined their 2003 estimates of maximum perennial yield for the Big Bear surface-water drainage basin on the basis of data collected since 2001. The revised estimates ranged from 2,510 to 2,585 acre-ft/yr. The greatest difference between the refined estimated perennial yield from 2006 and the 2003 estimates was a 150-230 acre-ft/yr reduction in estimated maximum perennial yield in the Mill Creek subbasin.

Baldwin Lake Surface-Water Drainage Basin

LeRoy Crandall & Associates (1987) and GeoScience Support Services, Inc. (1992, 1999) used water balance (Roscoe-Moss, 1990), Darcian flow (Roscoe-Moss, 1990), zero net draft (Chow, 1964), and Hill (Chow, 1964) methods to calculate maximum perennial yield for subbasins in the Baldwin Lake surface-water drainage basin. LeRoy Crandall & Associates (1987) estimated that the perennial yield for the Baldwin Lake surface-water drainage basin ranged from 1,300 to 1,330 acre-ft/yr. GeoScience Support Services, Inc. (1999) derived higher values, however, estimating that the perennial yield in the Baldwin Lake surface-water drainage basin ranged from 2,400 to 3,800 acre-ft/yr and that the total recoverable water (the sum of surface runoff and recharge to groundwater) for the Baldwin Lake surface-water drainage basin was about 4,900 acre-ft/yr.

Monthly Water-Balance Modeling: Basin Characterization Model

By Lorraine E. Flint and Alan L. Flint

The Basin Characterization Model (BCM; Flint and others, 2004) was refined and applied to the Big Bear Valley. The BCM uses a mathematical deterministic water-balance approach to estimate in-place recharge and runoff in a basin. The model uses the distribution of precipitation, snow accumulation and melt, potential evapotranspiration, soil-water storage, and bedrock permeability to estimate a monthly water balance for the groundwater system. A thirty-year normal record of monthly precipitation and air temperature for 1971–2000 was used in this study as model inputs to simulate basin recharge and runoff during varying climatic conditions.

Model Description

The BCM is used to identify locations and climatic conditions that result in excess water (precipitation minus potential evapotranspiration) in a basin. Depending on the soil and bedrock permeability, excess water is partitioned for each grid cell as either (1) in-place recharge or (2) runoff that can become recharge in the alluvial basin. The spatial and temporal distribution of net infiltration (water infiltrating below the soil root zone) is dependent on precipitation, soil-water storage, bedrock permeability, and evapotranspiration, all of which can be estimated with available data on a regional scale. Using this approach, the most probable locations for potential in-place recharge and runoff can be identified. Total potential recharge is the combination of in-place recharge and a proportion of the runoff that is assumed to become recharge. Although the percentage of runoff that becomes recharge in the Big Bear area is unknown, studies in basins throughout the Great Basin indicate a range of 10-80 percent (Flint and Flint, 2007a); about 10 percent of the runoff in the southern regions becomes recharge (Izbicki, 2002; Hevesi and others, 2003). Thus, for this study, it was assumed that 10 percent of the runoff becomes recharge.

The BCM incorporates spatially-distributed estimates of monthly precipitation, monthly minimum and maximum air temperature, monthly potential evapotranspiration, soil-water storage, and bedrock permeability at a spatial resolution matching that of the available digital elevation model, in this case 885-ft (270-m) grid cells derived from the 30-m Elevation Derivatives for National Applications map (EDNA; *http:// edna.usgs.gov*). Calculations to determine the components of the water balance were made to determine the area in a basin where excess water is available and whether or not it can be stored in the soil or infiltrate into the underlying bedrock at an estimated rate equivalent to the bedrock permeability. Potential evapotranspiration was partitioned on the basis of vegetation cover to represent bare-soil evaporation and evapotranspiration through vegetation.

Water-Balance Calculations

The BCM code (Flint and others, 2011) is written in FORTRAN-90 and uses ASCII files of distributed upper boundary conditions and surface properties as input to calculate potential recharge and runoff. A series of water-balance equations were developed to calculate the area and the amount of potential in-place recharge and runoff for each basin. For example, each 885-ft grid cell was analyzed each month to determine water availability for recharge. The available water (AW) for potential in-place recharge, potential runoff, or water to be carried over to the following month is defined as follows:

$$AW = P + S_m - PET - S_a + S_s \tag{1}$$

where

P is precipitation,

S_m is snowmelt,

PET is potential evapotranspiration,

S_a is snow accumulation and snow pack carried over from the previous month, and

S_s is stored soil water carried over from the previous month.

All units are in inches per month. Potential runoff was calculated as the available water minus the total storage capacity of the soil (soil porosity multiplied by soil depth). Potential in-place recharge was calculated as the available water remaining (after runoff) minus the field capacity of the soil (the water content at which drainage becomes negligible). Maximum in-place recharge on a unit grid-cell basis is the permeability of the bulk bedrock (cubic inches of water per square inch of grid-cell area per month). If the total soil-water storage is reached, the potential in-place recharge is equal to the bedrock permeability. Any water remaining after the monthly time step is carried over into the next month in the S_s term.

Model Input and Data Requirements

Soil-water storage capacity and soil-infiltration capacity were estimated using soil-texture estimates and permeability from the State Soil Geographic Database (STATSGO; http://www.ncgc.nrcs.usda.gov/products/datasets/statsgo/), a state-compiled geospatial database of soil properties that generally are consistent across state boundaries (U.S. Department of Agriculture-National Resource Conservation Service, 1994). Soil thickness was estimated using available geologic maps and the STATSGO database. The soil thickness was assumed to be 20 ft in all areas mapped as Quaternary alluvium (figs. 10 and 11) in the Big Bear Valley of the California

1:750,000 geologic map (Jennings, 1977). In all other areas, the STATSGO database was used to estimate soil depths from 0 to 5 ft (fig. 11). Uncertainties in soil properties are primarily due to the spatial resolution available in the STATSGO database, the weighted averaging used for calculating average properties for each mapped soil unit, and the estimations of hydraulic properties from soil textural classification. More details regarding uncertainties in soil properties are discussed by Hevesi and others (2003).

Saturated hydraulic conductivity of bedrock was estimated by using the California 1:750,000 geologic map (Jennings, 1977) and the estimated values for the different geologic units are presented on the map (fig. 10). Initial

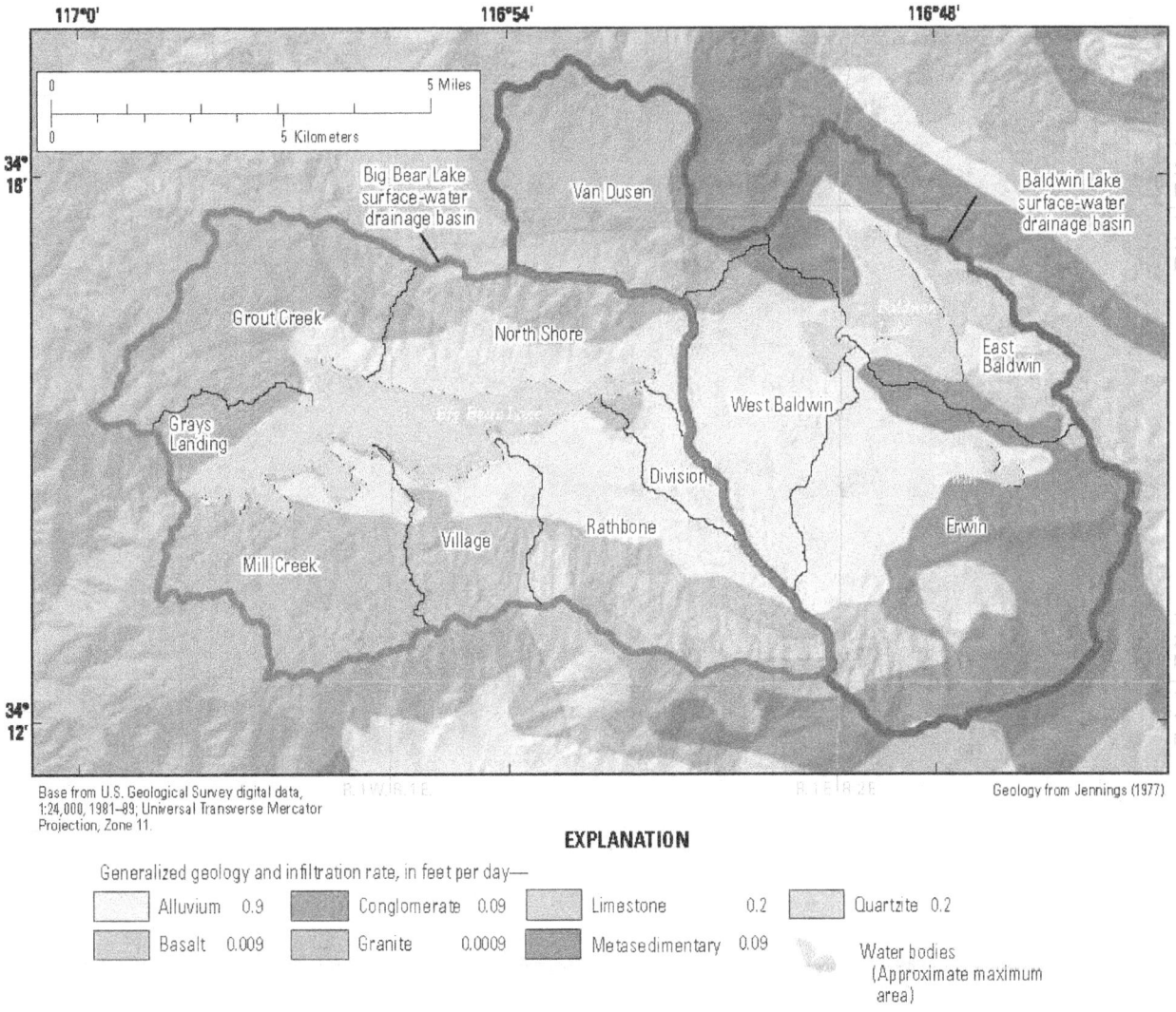

Base from U.S. Geological Survey digital data, 1:24,000, 1981–89; Universal Transverse Mercator Projection, Zone 11.

Geology from Jennings (1977)

EXPLANATION

Generalized geology and infiltration rate, in feet per day—

Alluvium 0.9	Conglomerate 0.09	Limestone 0.2	Quartzite 0.2
Basalt 0.009	Granite 0.0009	Metasedimentary 0.09	Water bodies (Approximate maximum area)

Figure 10. Generalized geology and saturated hydraulic conductivity of bedrock used in the Basin Characterization Model of Big Bear Valley, San Bernardino County, California.

estimates of saturated hydraulic conductivity of bedrock were based on literature values, aquifer-test results, surface-based infiltration experiments, and expert opinion from field geologists. The hydraulic properties of macropores and fractures are incorporated in the bulk estimates of hydraulic conductivity. Hydraulic-conductivity estimates of bedrock are uncertain because of the unknown hydraulic properties and spatial distributions of fractures, faults, fault gouge, and shallow infilling materials associated with different bedrock types and evaporative demand. Many of the values have been refined on the basis of model calibrations done in the Mojave Desert (Hevesi and others, 2003) and southern California (Rewis and others, 2006). The highest saturated hydraulic-conductivity values

were assigned to the alluvial deposits, at 0.9 feet per day (ft/d) and the highly fractured carbonates, at 0.2 ft/d; the lowest values, 9×10^{-4} ft/d, were assigned to granitic rock types). It is assumed infiltrating water that reaches depths of 20 ft will eventually become recharge. A depth of 20 ft was chosen on the basis of field observations of desert plant root penetration into alluvium and bedrock in the Mojave Desert and assumes that all processes controlling net infiltration are within the top 20 ft of the surficial materials, as shown by Flint and Flint (1995) for Yucca Mountain in the southern Great Basin.

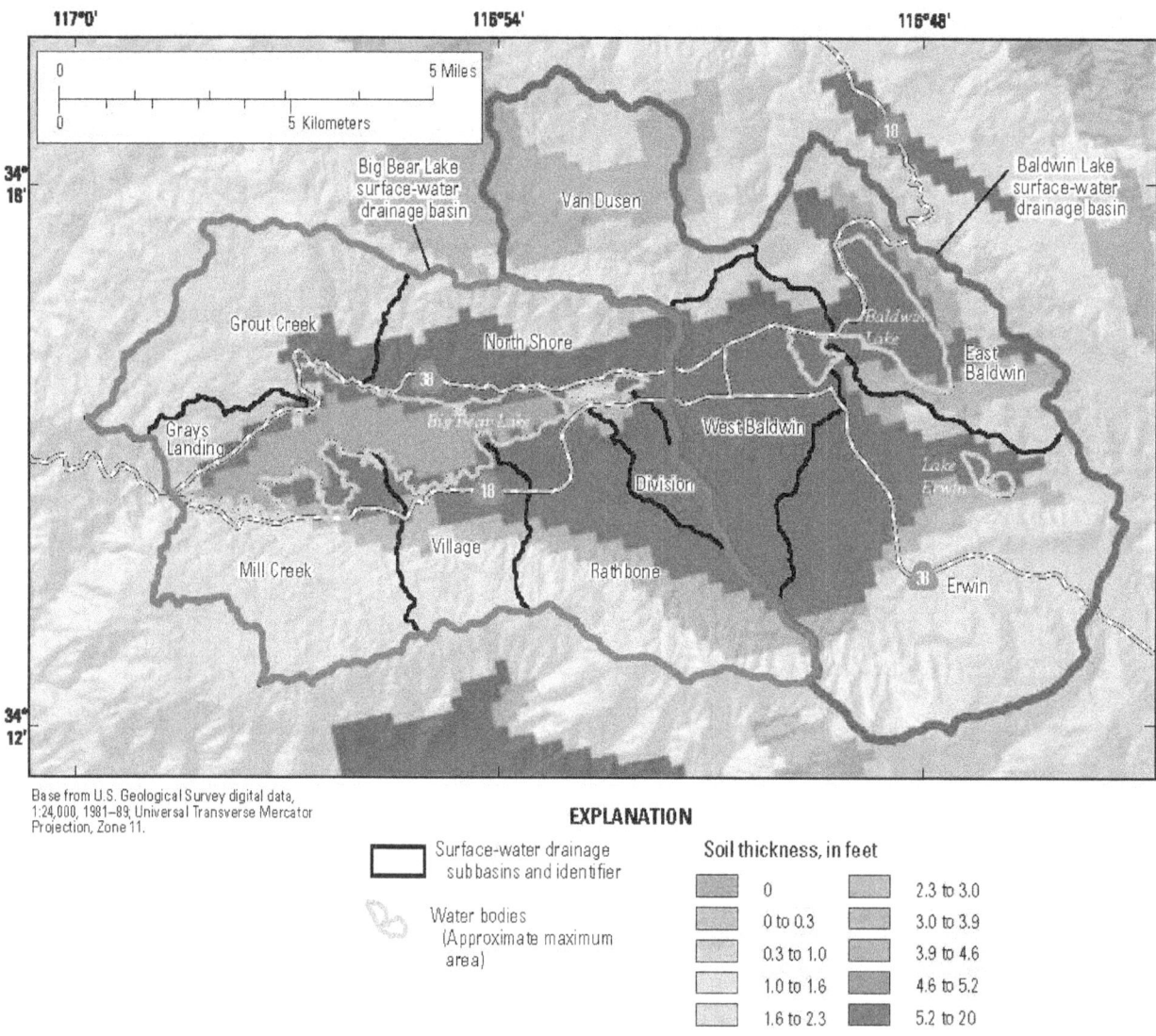

Base from U.S. Geological Survey digital data,
1:24,000, 1981–89, Universal Transverse Mercator
Projection, Zone 11.

EXPLANATION

☐ Surface-water drainage
   sub basins and identifier

🌀 Water bodies
   (Approximate maximum
   area)

Soil thickness, in feet

0	2.3 to 3.0
0 to 0.3	3.0 to 3.9
0.3 to 1.0	3.9 to 4.6
1.0 to 1.6	4.6 to 5.2
1.6 to 2.3	5.2 to 20

Figure 11.   Generalized soil thickness used in the Basin Characterization Model of Big Bear Valley, San Bernardino County, California.

Climate (air temperature and precipitation) was simulated in the BCM by using data from the 30-yr period, 1971–2000. The precipitation (*fig. 2*) and temperature data for this period are available as monthly averages from the Parameter Regression on Independent Slopes Model (PRISM) at approximately 2.5-mi grid spacing (Daly and others, 2004). These data were downscaled to the 885-ft grid cells used for this study by using a model from Nalder and Wein (1998) that combines a spatial gradient plus inverse distance squared weighting (GIDS) with monthly point data to interpolate to each grid cell using multiple regression. Parameter weighting is based on location and elevation following the equation:

$$Z = \frac{\left[\sum_{i=1}^{N} \dfrac{Z_i + (X - X_i) \times C_x + (Y - Y_i) \times C_y + (E - E_i) \times C_e}{d_i^2} \right]}{\left[\sum_{i=1}^{N} \dfrac{1}{d_i^2} \right]} \quad (2)$$

where

Z = estimated climatic variable, at a specific 885-ft grid cell defined by X, Y, and E, which are easting, northing, and elevation, respectively;

Z_i = value of PRISM cell i;

X_i, Y_i, E_i = easting, northing, elevation of PRISM cell i;

N = number of PRISM cells;

d_i = distance from the site to PRISM cell i; and

C_x, C_y, C_e = regression coefficients for easting,

A search radius of about 6 mi was used to limit the influence of distant data. For each 885-ft grid-cell estimate of temperature or precipitation, approximately 25 PRISM grid cells were used to estimate the regression coefficients (C_x, C_y, C_e).

A computer program created by Flint and Childs (1987) was modified to estimate potential evapotranspiration. The modified program calculates solar radiation for each grid cell in the model domain on the basis of percentage of sky viewed because of topographic shading; when combined with air temperature, the solar radiation is converted to net radiation and soil heat flux (Shuttleworth, 1993). The result was used with the Priestley–Taylor equation (Priestley and Taylor, 1972) to estimate potential evapotranspiration (*fig. 12*), which was corrected for vegetated and bare-soil area using estimates of vegetation cover from vegetation maps (National Gap Analysis Program; *http://www.gap.uidaho.edu*). The regional-scale approach used with the BCM assumes that potential evapotranspiration can be used to provide an estimate of potential recharge that is a lower bound for the purpose of evaluating the mechanisms controlling recharge, runoff, and the differences between basins.

Snow accumulation and ablation was simulated by using an adaptation of the operational National Weather Service (NWS) energy and mass balance model, the Snow-17 model as described by Anderson (1976) and Shamir and Georgakakos (2005). The model was used to calculate the potential for snowmelt as a function of air temperature and an empirical snowmelt factor that varies with the day of the year (Lundquist and Flint, 2006). Snow depth was calculated for areas where precipitation occurs and air temperature is 34.7°F or below. Sublimation of snow was calculated as a percentage of potential evapotranspiration. Calibration of snow accumulation and snowmelt was completed using MODIS snow cover remote sensing data (MODIS/Aqua snow cover 8-day L3 Global 500-m grid, version 4; *http://nsidc.org/data*; ordered June 2004) for comparison. The snow accumulation and melt coefficients were adjusted iteratively to optimize the fit of simulated snow cover to the measured MODIS snow cover by varying the temperature threshold at which accumulation and melt occurs (Lundquist and Flint, 2006). Examples comparing satellite data with modeled snow cover are shown for January 2001 (*fig. 13A*), when snow cover is at its approximate maximum, and for March 2001 (*fig. 13B*), when snowmelt processes are dominant. As shown in *figures 13A* and *B*, the area of modeled snow cover reasonably matches the area of measured data. The BCM model allows snow pack and soil moisture to be carried over in the calculations from month to month, which becomes important when temperatures are cold enough for precipitation to form snow. Because snow can persist for several months before melting, large volumes of water can be made available for potential recharge in a single monthly model time step.

BCM Simulated In-Place Recharge and Runoff for Water Years 1971–2000

To estimate natural groundwater recharge of the Big Bear Valley, monthly in-place recharge and runoff were simulated using the BCM for water years 1971–2000 (*fig. 14*). The BCM simulated approximately 12,700 acre-ft of potential in-place recharge and 30,800 acre-ft of potential runoff in the Big Bear and Baldwin Lakes surface-water drainage basins. Most of the simulated in-place recharge occurs at high elevations in the southern part of the Baldwin Lake surface-water drainage basin and along the ridges that form the northern edge of the Big Bear Lake and Baldwin Lake surface-water drainage basins. The locations where simulated runoff is the highest are along the entire perimeter of Big Bear Lake surface-water drainage basin, where precipitation is highest (*fig. 2*).

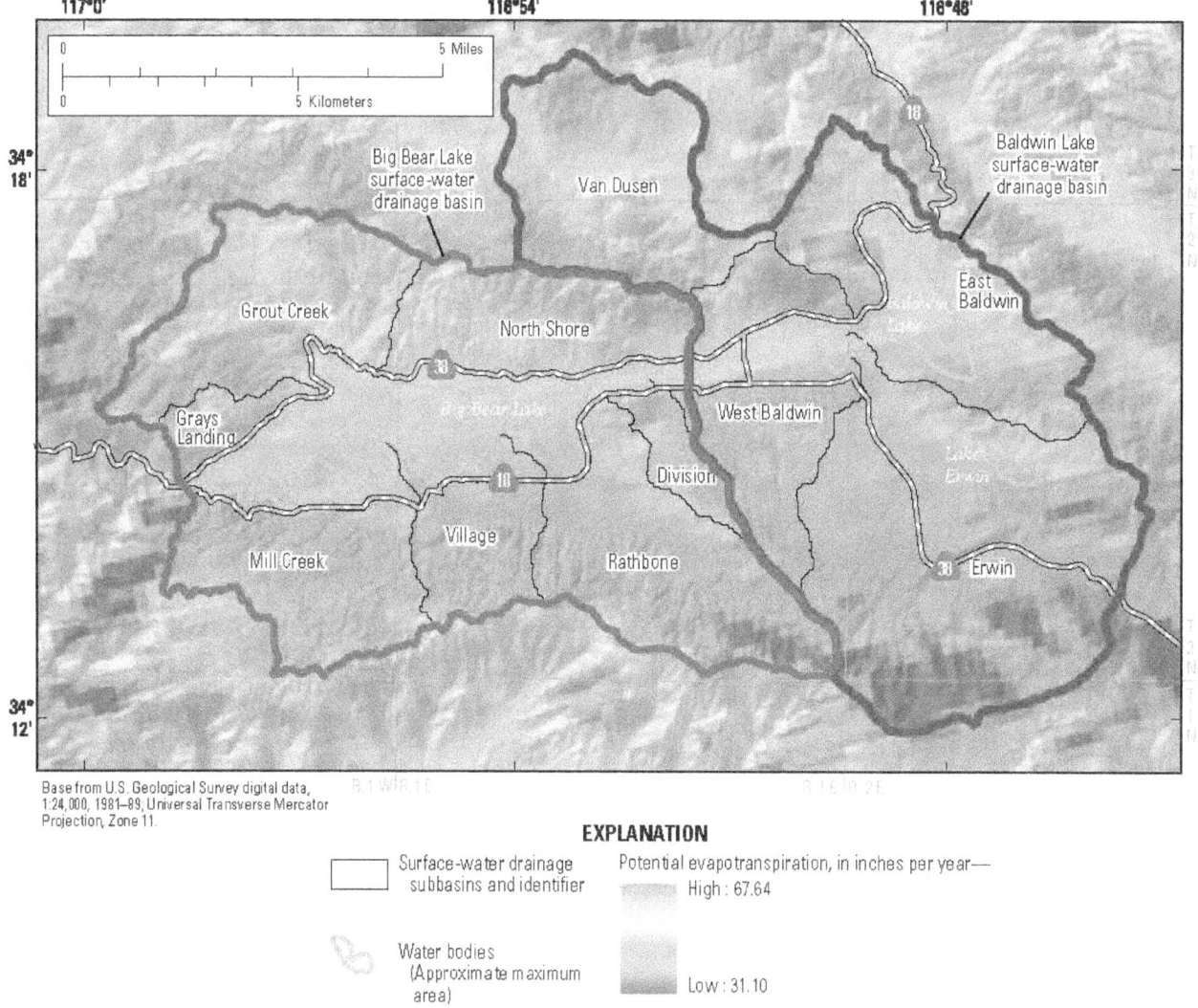

Figure 12. Modeled potential evapotranspiration used in the Basin Characterization Model of Big Bear Valley, San Bernardino County, California.

Assuming that 10 percent of the runoff becomes recharge, approximately 15,800 acre-ft of potential recharge (simulated in-place recharge plus 10 percent of simulated runoff) is simulated to occur within Big Bear Valley; about 6,600 acre-ft/yr in the Big Bear Lake surface-water drainage basin and about 9,200 acre-ft/yr in the Baldwin Lake surface-water drainage basin (*table 3*). The actual percentage of runoff that becomes recharge in Big Bear Valley is not known but could be determined in subsequent studies and analyses. Flint and others (2004), in their study of potential recharge

in Nevada, assumed that 10 percent of the BCM simulated runoff becomes recharge. They determined that this value ranged from 10 percent in southern Nevada to 80 percent in some locations in northern Nevada. The spatial distribution of total potential recharge using the BCM approach indicates that the greatest amount of recharge occurs in the mountains that surround the Big Bear Lake and Baldwin Lake surface-water drainage basins, with smaller amounts of recharge occurring on the valley floor in the Baldwin Lake surface-water drainage basin (*fig. 15A*).

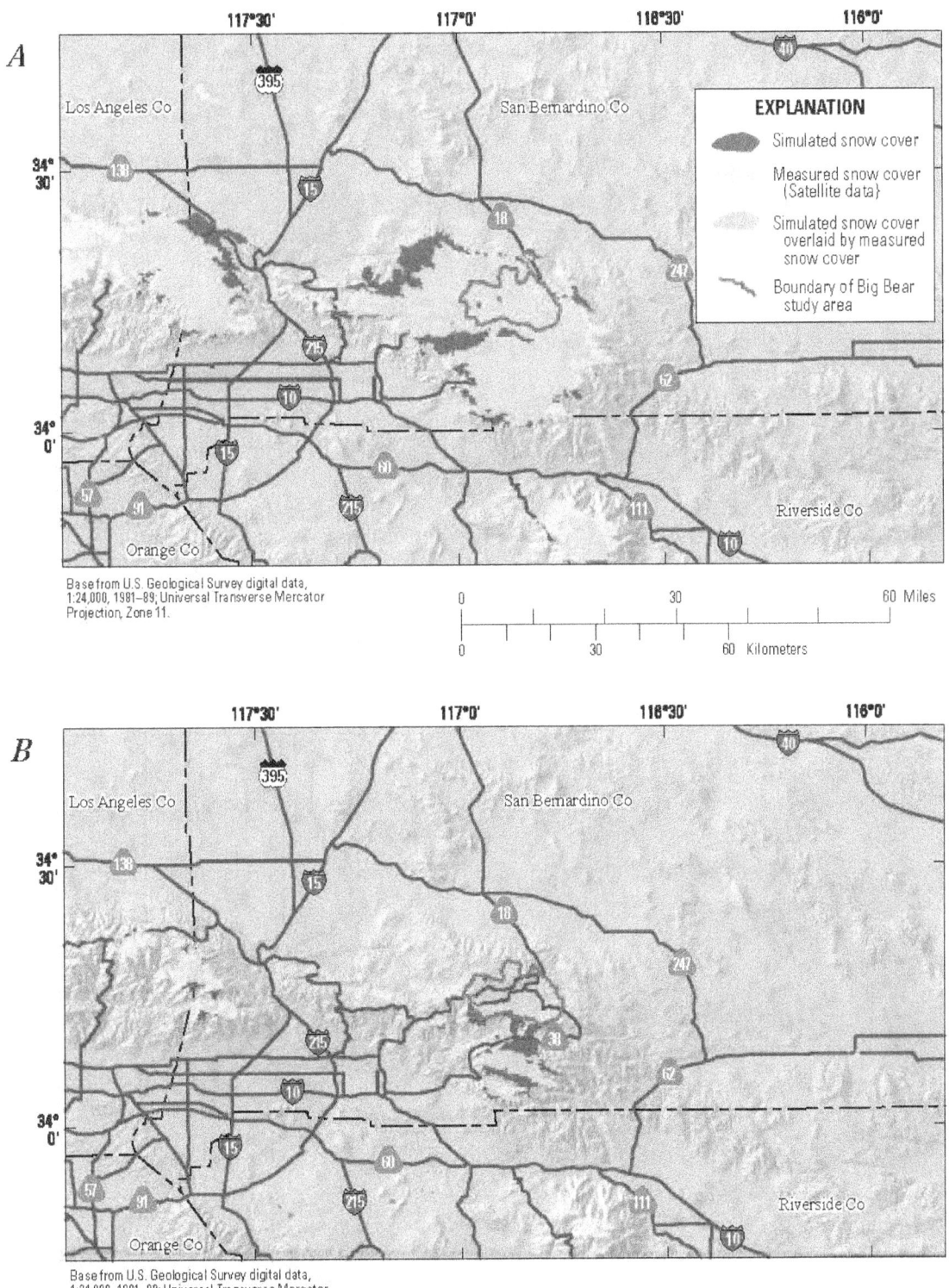

Figure 13. Comparison of snow cover estimated from the Basin Characterization Model and from MODIS satellite data for (A) January 2001 and (B) March 2001 for Big Bear Valley, San Bernardino County, California.

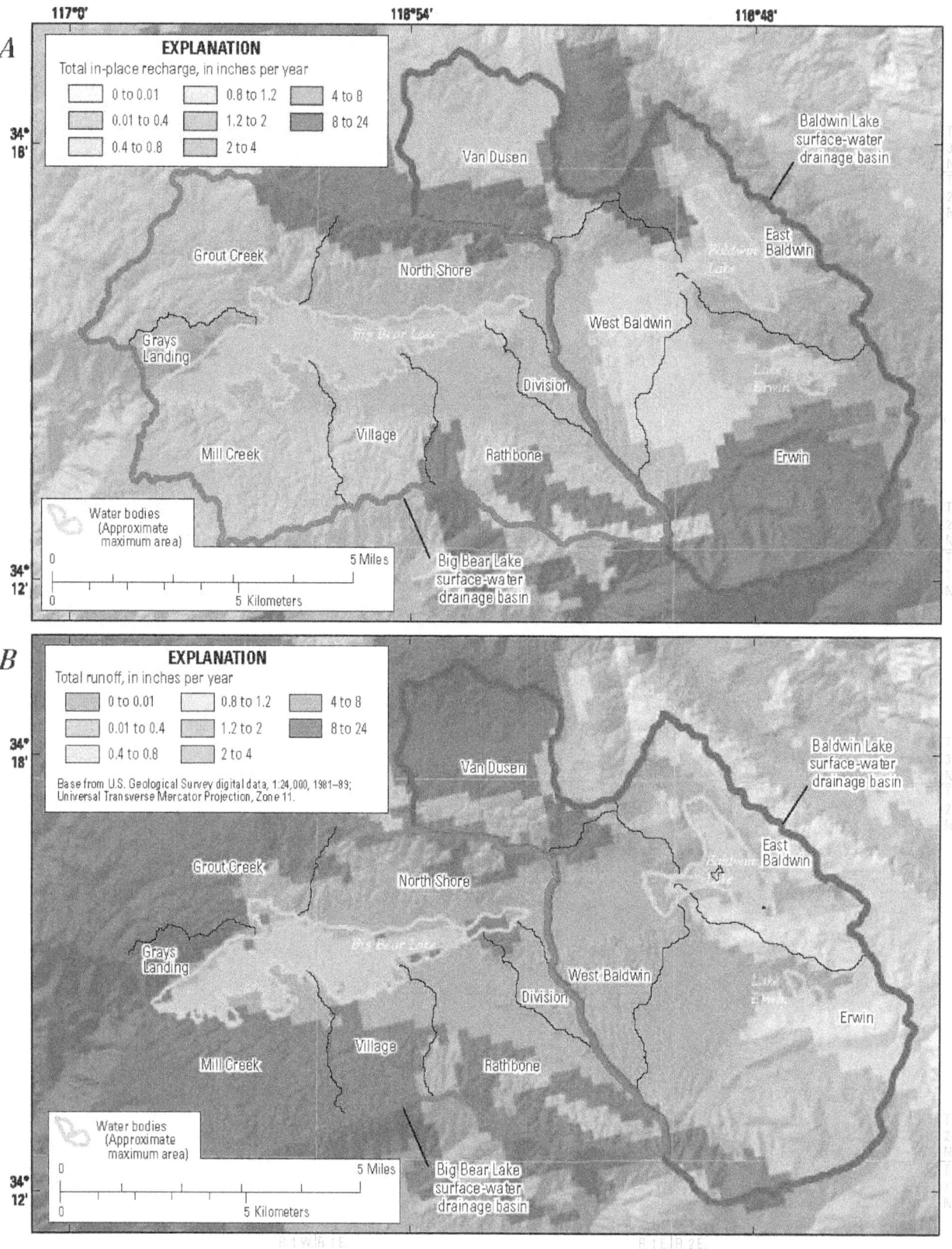

Figure 14. (A) Total in-place recharge and (B) total runoff estimated using the Basin Characterization Model for Big Bear Valley, San Bernardino County, California.

Table 3. Precipitation, evapotranspiration, runoff, and recharge estimated by using the Basin Characterization Model for water years 1971–2000 in Big Bear Valley, San Bernardino County, California.

[Measurements are yearly averages. Total potential recharge is in-place recharge + 10 percent runoff. **Abbreviations:** N/A, not applicable; –, no data available]

Subbasin	Precipitation (inches)	Precipitation (acre-feet)	Potential evapotranspiration (inches)	Potential evapotranspiration (acre-feet)	In-place recharge (acre-feet)	Runoff (acre-feet)	Total potential recharge for whole basin (inches)	Total potential recharge for whole basin (acre-feet)	Dominant hydrologic response
			Big Bear Lake surface-water drainage basin						
Big Bear Lake	25.4	886	55.2	1,924	506	4,941	3.7	1,000	–
Grout Creek	28.3	10,131	5.5	19,874	461	5,701	2.9	1,032	Runoff
Mill Creek	29.4	11,738	52.4	20,920	92	6,163	2.1	709	Runoff
Village	26.3	10,002	51.6	19,601	130	2,187	2	349	Runoff
Rathbone	26.8	3,114	50.9	5,912	1,705	2,926	4.8	1,998	Runoff
Division	23.6	4,025	53.6	9,157	207	201	1.7	227	Runoff
North Shore	23.8	5,554	58.3	13,605	1,060	1,119	4.1	1,172	Runoff
Gray's Landing	29.3	1,682	58.6	3,367	33	1,120	2.4	145	Runoff
Average/total	27	46,124	53.9	92,051	4,196	24,359	3.2	6,632	N/A
			Baldwin Lake surface-water drainage basin						
Van Dusen	25.5	9,220	57.1	20,681	1,107	4,337	4.2	1,540	Runoff
West Baldwin	23.2	7,788	55.8	18,711	369	199	1.6	389	Runoff
Erwin	27.6	19,606	49.9	35,453	5,272	1,668	5.5	5,439	Recharge
East Baldwin	21.8	8,803	55.7	22,474	1,771	258	3.3	1,797	Recharge
Average/total	25.3	45,927	53.4	96,783	8,519	6,462	4.1	9,166	N/A
Study area average/total	26.2	92,051	53.7	188,834	12,715	30,821	3.7	15,796	N/A

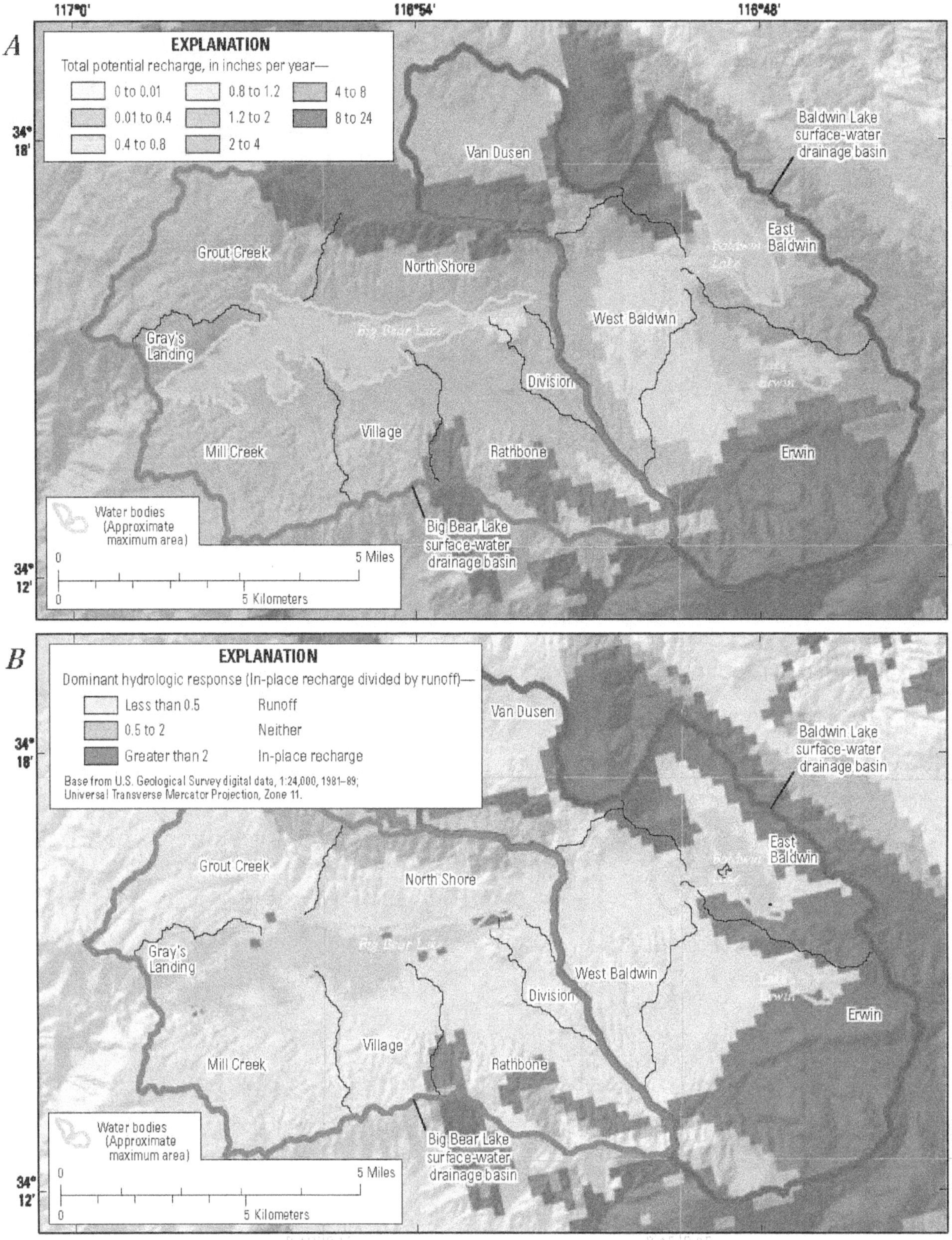

Figure 15. (*A*) Total potential recharge calculated as in-place recharge plus 10 percent of runoff and (*B*) dominant hydrologic response calculated as recharge divided by runoff using the Basin Characterization Model for Big Bear Valley, San Bernardino County, California.

The ratio of simulated in-place recharge to runoff is an indicator of the dominant hydrologic response at a given location: a ratio of 0 to 0.5 indicates that runoff is dominant, with volumes at least twice that of in-place recharge; a ratio 0.5 to 2.0 indicates that neither hydraulic response is dominant; and a ratio greater than 2.0 indicates that in-place recharge is dominant, with volumes at least twice that of runoff. Most of the study area is dominated by runoff (*fig. 15B*). In-place recharge typically dominates in areas having high-permeability carbonate rocks; in contrast, runoff typically dominates in areas where there are low-permeability granitic and metamorphic rocks. Whereas most of Big Bear Lake surface-water drainage basin is dominated by runoff, which maintains the large man-made lake, Baldwin Lake surface-water drainage basin is dominated by in-place recharge in all the high elevation locations (*figs. 1, 15B*), resulting in less runoff in the basin—likely accounting for the intermittent presence of a large water body.

Simulated annual in-place recharge and runoff were compared to precipitation for the two major precipitation indices (*fig. 16*), Pacific Decadal Oscillation (PDO) and El Nino Southern Oscillation (ENSO). The ENSO index appears to be correlated better to the precipitation, simulated in-place recharge, and simulated runoff and recharge in the Big Bear and Baldwin Lake surface-water drainage basins than the PDO index. Many of the simulated low in-place recharge and runoff years correspond to negative ENSO years, when there is almost no simulated runoff in Baldwin Lake surface-water drainage basin. Dry periods in the 1970s and post-1998 occurred during a negative PDO; however, the severe drought in 1984–90 occurred in a positive PDO. The relation of simulated recharge to the ENSO climate cycle, especially when ENSO is negative, could be used to project future recharge in the Big Bear Valley.

Long-Term Recharge

Long-term recharge for the Big Bear area was estimated by relating annual potential recharge simulated using BCM (in-place recharge plus 10 percent of runoff) to annual precipitation for the simulation period, 1971–2000 (*fig. 17A*), and then using this relation to estimate annual potential recharge for the long-term precipitation record, 1895–2004, compiled by Daly and others (2004; *fig. 17B*). The potential recharge values in 1969 and 1978 (*fig. 17B*) were the highest for the last century. The 5-year running average of the precipitation and potential recharge (*fig. 17B*) indicates an increase in the variability of precipitation and potential recharge since the late 1960s. This variability is more pronounced in the potential recharge record because of the non-linear response of potential recharge to precipitation. When the precipitation and corresponding potential recharge estimates are compared to changes in measured water levels in local wells for the 1986–2005 period (*fig. 17C*), the variability in water levels coincides with the variability in the precipitation and potential-recharge records defined by the 5-year running average of these records. Yearly variations among precipitation, potential-recharge, and water-level records are present, although they are less discernable because of the variation in pumping. This relation likely has occurred for the last century and shows the sensitivity of the local groundwater system to changes in climate. The relation among precipitation, potential recharge, and water levels indicates that water availability is sensitive to yearly climate fluctuations, as well as long-term fluctuations in climate, which is described in detail and supported in Flint and Flint (2007a). Assuming these relations remain into the future, the responses to future changes in climate are likely to be similar.

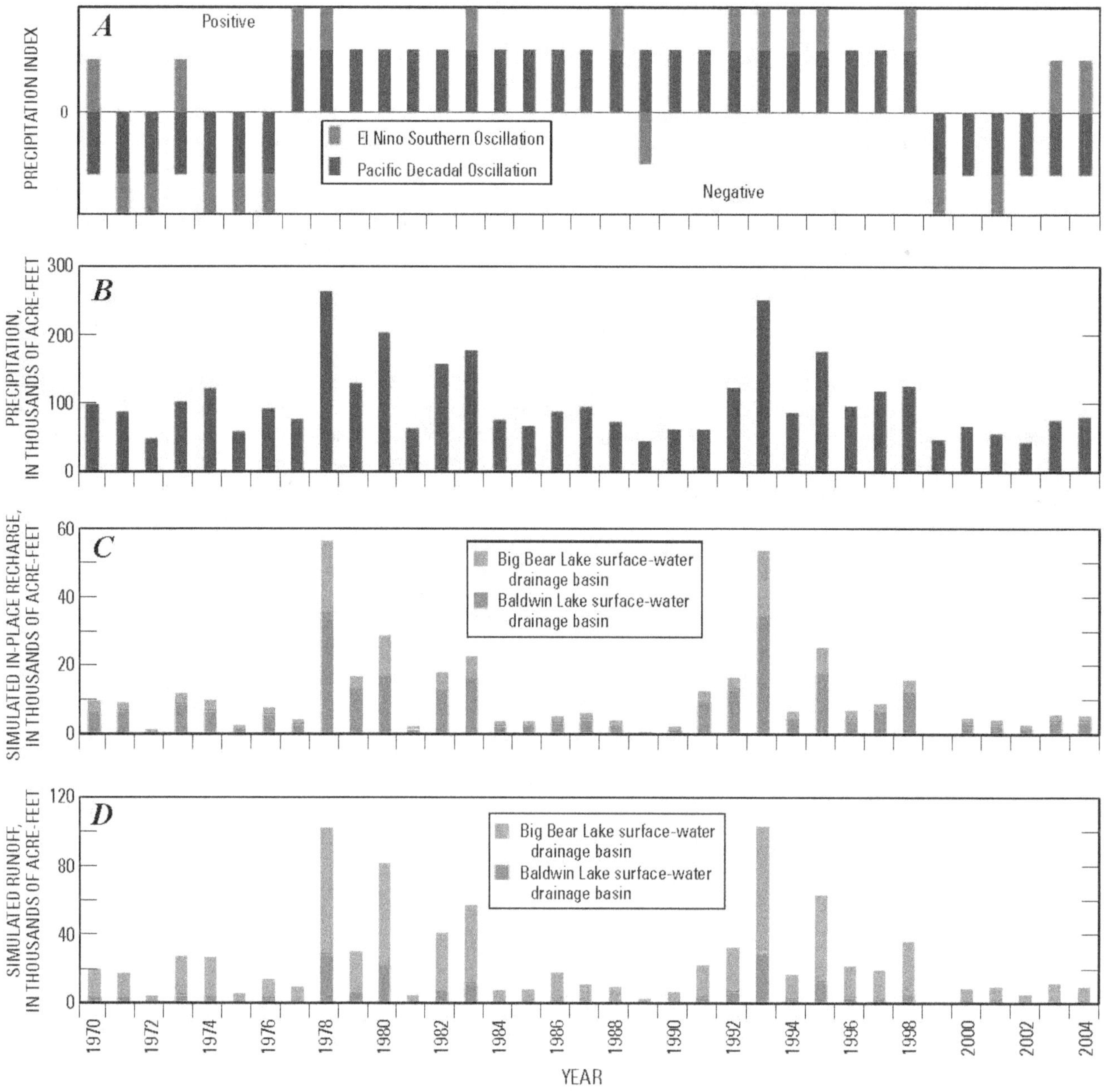

Figure 16.   Annual time series of (*A*) precipitation indices, Pacific Decadal Oscillation and El Nino Southern Oscillation, (*B*) precipitation, (*C*) Basin Characteristic Model (BCM) simulated in-place recharge, and (*D*) BCM simulated runoff for Big Bear Valley, San Bernardino County, California, 1970-2004.

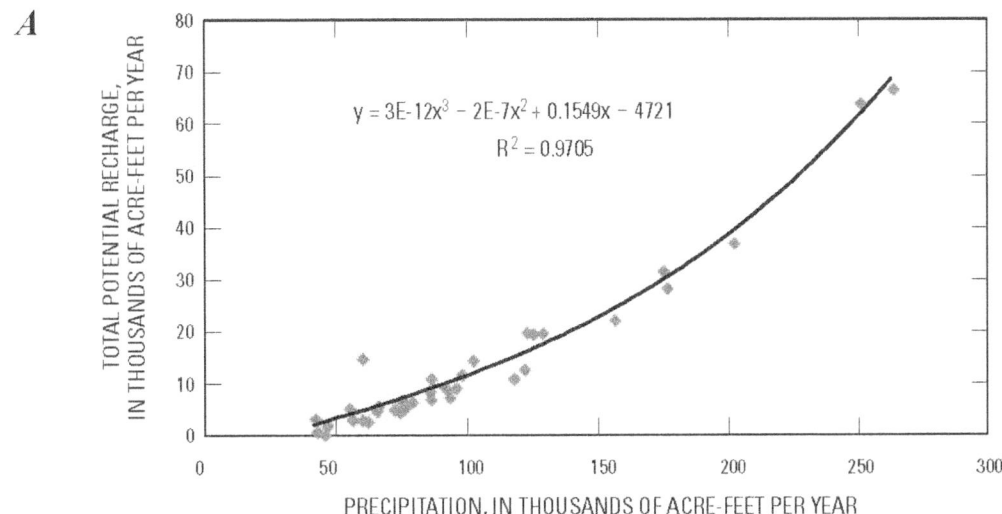

A

$$y = 3E\text{-}12x^3 - 2E\text{-}7x^2 + 0.1549x - 4721$$

$$R^2 = 0.9705$$

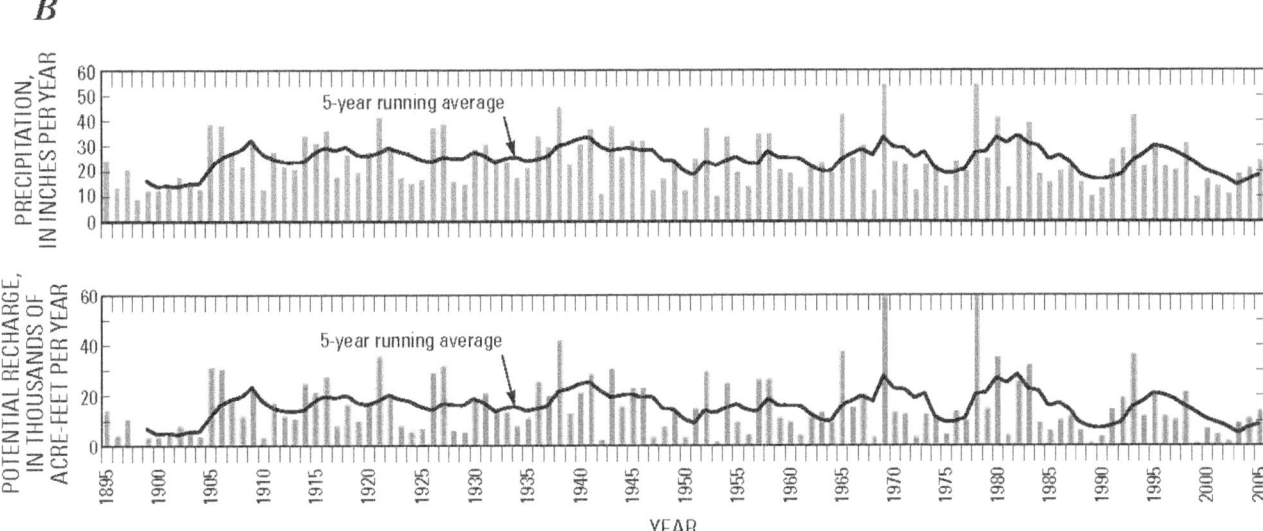

B

Figure 17. (*A*) Relation between annual precipitation and Basin Characteristic Model (BCM) simulated potential recharge for 1971–2000, (*B*) long-term annual (1895–2005) potential recharge estimated from measured precipitation, and (*C*) comparison of annual potential recharge to measured water levels in the Big Bear Valley, San Bernardino County, California, 1986–2005.

C

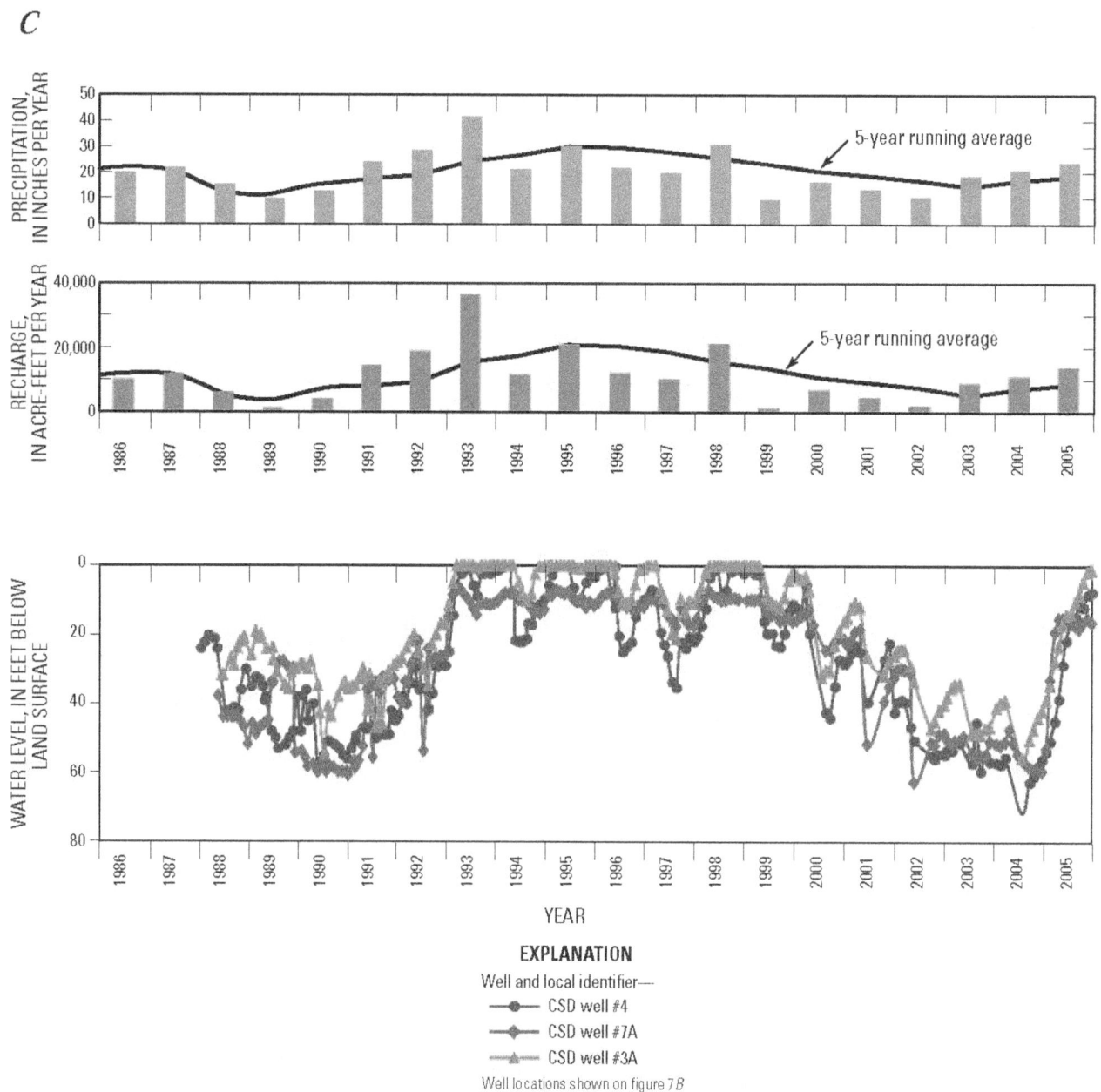

Figure 17.   Continued

Daily Rainfall-Runoff Modeling: INFILv3

By Joseph A. Hevesi

A daily rainfall-runoff model, INFILv3, was used to develop estimates of natural groundwater recharge for the Big Bear Valley. The model was calibrated to available records of lake levels and volumes for Big Bear and Baldwin Lakes. A high-resolution spatial discretization of the study area was used to incorporate detailed geology and vegetation maps, which were used to define the physical characteristics of the Big Bear Valley. Daily climate data developed from a local network of monitoring sites and published average annual precipitation maps were used to develop inputs for the INFILv3 model.

The INFILv3 model has been previously applied to studies of groundwater recharge in the southern California region, including the Death Valley regional flow system (Hevesi and others, 2003), the Joshua Tree area (Nishikawa and others, 2004), and the San Gorgonio Pass area (Rewis and others, 2006). In these studies, the INFILv3 model provided an estimate of recharge based on simulated daily net infiltration, where net infiltration is defined as the percolation of water from rain, snowmelt, and streamflow below the maximum depth of the root zone or the zone of evapotranspiration (Hevesi and others, 2003). Daily net infiltration and evapotranspiration are simulated by INFILv3 by using a multi-layered representation of the root-zone, and simulated daily runoff is allowed to infiltrate back into the root-zone during the process of flow routing, thereby accounting for the effects of streamflow on recharge (fig. 18A).

As indicated in the original version of INFILv3, net infiltration in a groundwater basin is not necessarily equivalent to recharge in that basin because water that infiltrates past the root zone does not always reach the water table (Hevesi and others, 2003). The potential for differences between simulated net infiltration and actual groundwater recharge tends to increase with increases in unsaturated-zone thickness, travel time of the infiltrated water through the unsaturated zone, climate variability, and geologic heterogeneity in the unsaturated zone (Flint and others, 2000). In mountainous areas, the unsaturated zone is likely to be more geologically heterogeneous, increasing the potential for localized perching and lateral groundwater flow in the shallow subsurface. Lateral groundwater flow in the shallow subsurface, referred to in this study as seepage, can divert a portion of net infiltration downstream to springs or to subsurface locations within the zone of evapotranspiration. This might be especially true for steep mountain drainage basins underlain by low-permeability bedrock, such as the higher elevation areas of the Big Bear and Baldwin Lake surface-water drainage basins where springs are present. The net effect of seepage on a basin-wide scale is a decrease in recharge balanced by an increase in evapotranspiration and streamflow. In previous applications of INFILv3 in southern California (Rewis and others, 2006; Nishikawa and others, 2004), the seepage component of the water balance was not taken into account. As a result, the net infiltration that was simulated for tributary drainages in mountainous terrain overestimated the recharge necessary to calibrate groundwater models for downstream groundwater basins.

INFILv3 was modified for this study to include a perched zone beneath the root zone (fig. 18B) to better simulate seepage and, ultimately, recharge in the shallow subsurface in mountainous terrain. Simulating lateral seepage beneath the root zone and above the basin-wide water table allowed some of the shallow groundwater that originates as net infiltration to percolate downstream and discharge back into the root zone, contributing to evapotranspiration, runoff, and net infiltration at the downstream location. Shallow groundwater that does not contribute to seepage is stored in the perched zone and is available for recharge. Bedrock directly underlying the root zone is assumed to be more permeable than the bedrock at greater depth below the ground surface as a result of weathering and higher near-surface fracture densities. For a given rock type, the lateral hydraulic conductivity within this more permeable zone is assumed to be higher than the vertical hydraulic conductivity deeper within the bedrock, allowing the formation of a perched aquifer. In the case of unconsolidated deposits underlying the root zone (for example, locations with thick alluvium or basin-fill), the perched zone represents a perched aquifer lying on restricting layers below the root zone or along the contact between unconsolidated deposits and the underlying consolidated rock units.

The modified INFILv3 model simulates daily change in the amount of water stored in the perched zone as a function of net infiltration inflow from the root zone above, lateral-seepage outflow to the downstream grid cell, recharge outflow to the deeper bedrock or aquifer system, and the available storage capacity of the perched zone. Lateral seepage in the daily water balance allows for the representation of a baseflow component in simulated daily runoff that is caused by the delay between the time water infiltrates into the perched zone and the resulting lateral seepage that flows to downstream cells. Simulated seepage is returned to the root zone (layer 1) of downstream cells where the water can be lost to evapotranspiration, become streamflow, or become recharge in these downstream cells.

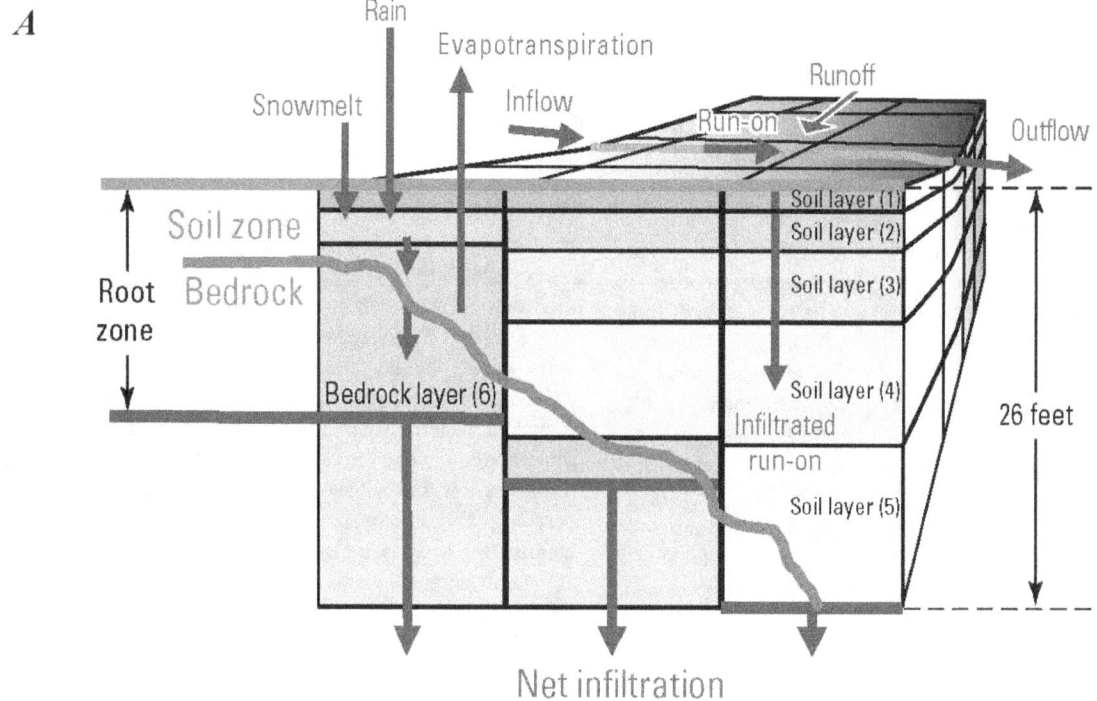

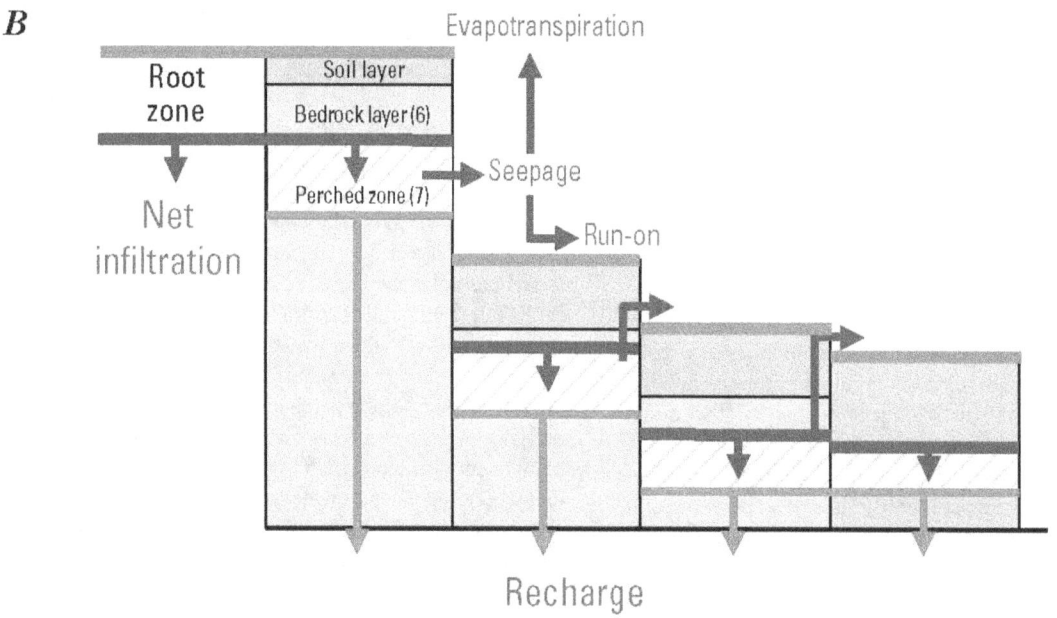

Figure 18.   Schematic showing (*A*) the multi-layered root zone simulated by INFILv3 and (*B*) the perched zone simulated by the modified version of INFILv3 used for Big Bear Valley, San Bernardino County, California.

The simulation of storage, seepage, and recharge from a perched zone in the modified INFILv3 model is consistent with approaches used in precipitation-runoff models such as Hydrologic Simulation Program–FORTRAN (HSPF; Bicknell and others, 1997) and Precipitation-Runoff Modeling System (PRMS; Leavesley and others, 1983). In these models, a shallow groundwater-storage component is used to model baseflow or throughflow contributions to total streamflow, and is represented as being separate from a deeper groundwater reservoir used to account for recharge. In the application of these models, simulated baseflow is an important component of total simulated streamflow and is usually a critical aspect of calibrating models to observed streamflow.

Model Description

The INFILv3 model uses a grid-based horizontal discretization of the drainage basin and a vertical discretization of the root zone. The root zone includes five layers that represent an upper soil component and a sixth layer that represents a lower geologic unit (either bedrock or unconsolidated deposits; fig. 18A). All root zone layers have variable thicknesses and are parameterized using maps of geology, soils, and vegetation. The bottom of the root zone is the estimated maximum depth below ground surface affected by evapotranspiration. The INFILv3 model was modified for this study to include a seventh layer that represents a perched zone beneath the root zone (fig. 18B). Net infiltration from the bottom of the root zone becomes inflow to the perched zone. Lateral seepage to downstream cells and recharge through the bottom of the perched zone are simulated as functions of the perched zone water content. The INFILv3 model does not directly account for interception storage and surface-retention storage; however, the model can indirectly account for these components by increasing the estimated soil thickness, which has the effect of increasing evapotranspiration. The water-balance calculations are based on water volumes rather than water mass because it is assumed that temperature effects on water density are negligible. The calculations use water-equivalent depths because all grid cells have equivalent areas. A detailed description of the original INFILv3 model is provided in Hevesi and others (2003), and documentation of the model is available on *http://water.usgs.gov/nrp/gwsoftware/Infil/Infil.html* (U.S. Geological Survey, 2008).

The INFILv3 model uses a daily time step for simulating the water balance of the root zone and perched zone. The simulated daily water balance of the root zone includes precipitation (as either rain or snow), snow accumulation, sublimation, snowmelt, infiltration into the root zone, evapotranspiration, percolation through the root zone, water-content changes for each root-zone layer, surface-water runoff, and net infiltration from the root zone (defined as drainage from the bottom root-zone layer; fig. 18). Potential daily evapotranspiration is simulated using an hourly time step to better represent the shading effects of rugged terrain relative to changes in solar position throughout the year. Daily evapotranspiration is simulated as a function of daily potential evapotranspiration and the vertical distribution of available water in the root zone layers.

The perched zone (layer 7) is assigned an upper hydraulic conductivity that defines the rate at which water vertically enters the top of the perched zone and a lower hydraulic conductivity that defines the maximum rate at which water vertically leaves the bottom of the perched zone as recharge. The upper hydraulic conductivity also is used to define the maximum lateral-seepage rate from upstream cells into downstream cells. The lateral-seepage rate is a function of a hydraulic gradient between two adjacent grid cells, the relative water content of each cell, and the upper hydraulic conductivity of each cell. A multiplier also is included to allow scaling of the upper hydraulic conductivity as a means of representing anisotropy or preferential lateral flow in the perched aquifer (this commonly is done to model preferential flow in watershed models). The hydraulic gradient is calculated using the elevation difference and horizontal distance between adjacent grid-cell centroids. The relative water content is calculated as the ratio of water stored in a grid cell at each time step to an assumed perched zone storage capacity of 2 ft (over the area of each grid cell), equal to 0.44 acre-ft per grid cell. If the perched zone storage capacity is exceeded for a given daily time step, the excess water is added back to the root zone; if the root zone is fully saturated, the excess water is added to the surface-water runoff.

Model parameters defining the root zone and the perched zone are spatially distributed across the study area using a horizontally discretized model grid. Input parameters defining the properties of the root zone and perched zone are uniquely defined for all model cells used to represent the spatially-varying physical characteristics of the drainage basin. The seepage flow network is defined using the two surface-water flow routing parameters that identify the upstream and downstream cell locations for each grid cell, based on standard convergent-flow routing methods (Hevesi and others, 2003; Maidment, 2002).

Model Discretization and Delineation of Model Areas

The INFILv3 model area and grid were developed using a 98.4-ft (30-m) digital elevation model (DEM) and the ARC-Hydro extension of ARC-GIS v9.0 (Maidment, 2002). The INFILv3 model area covers 72.1 mi[2] (46,122 acres), and is equivalent to the area of Big Bear Valley used in the BCM analysis. The model area was defined as the total area upstream of Bear Valley Dam, including both Big Bear Lake and Baldwin Lake surface-water drainage basins (*fig. 19*). The Baldwin Lake surface-water drainage basin is topographically upstream of the Big Bear Lake surface-water drainage basin; however, there is no historical evidence that overflows from

Baldwin Lake surface-water drainage basin have occurred into Big Bear Lake surface-water drainage (although little is known about groundwater transfers between basins). In this study, the Baldwin Lake surface-water drainage basin was modeled as a closed basin with respect to both surface-water and groundwater flow.

ARC-Hydro was used to define the streamlines, sub-drainage (model unit) boundaries, and drainage networks upstream of Bear Valley Dam. A modified 98.4-ft (30-m) DEM was developed in ARC-Hydro using the high-resolution 32.8-ft (10-m) National Hydrography Dataset (NHD; *http://nhd.usgs.gov*, accessed January 2007). The areas of Big Bear Lake and Baldwin Lake also were defined using the NHD data.

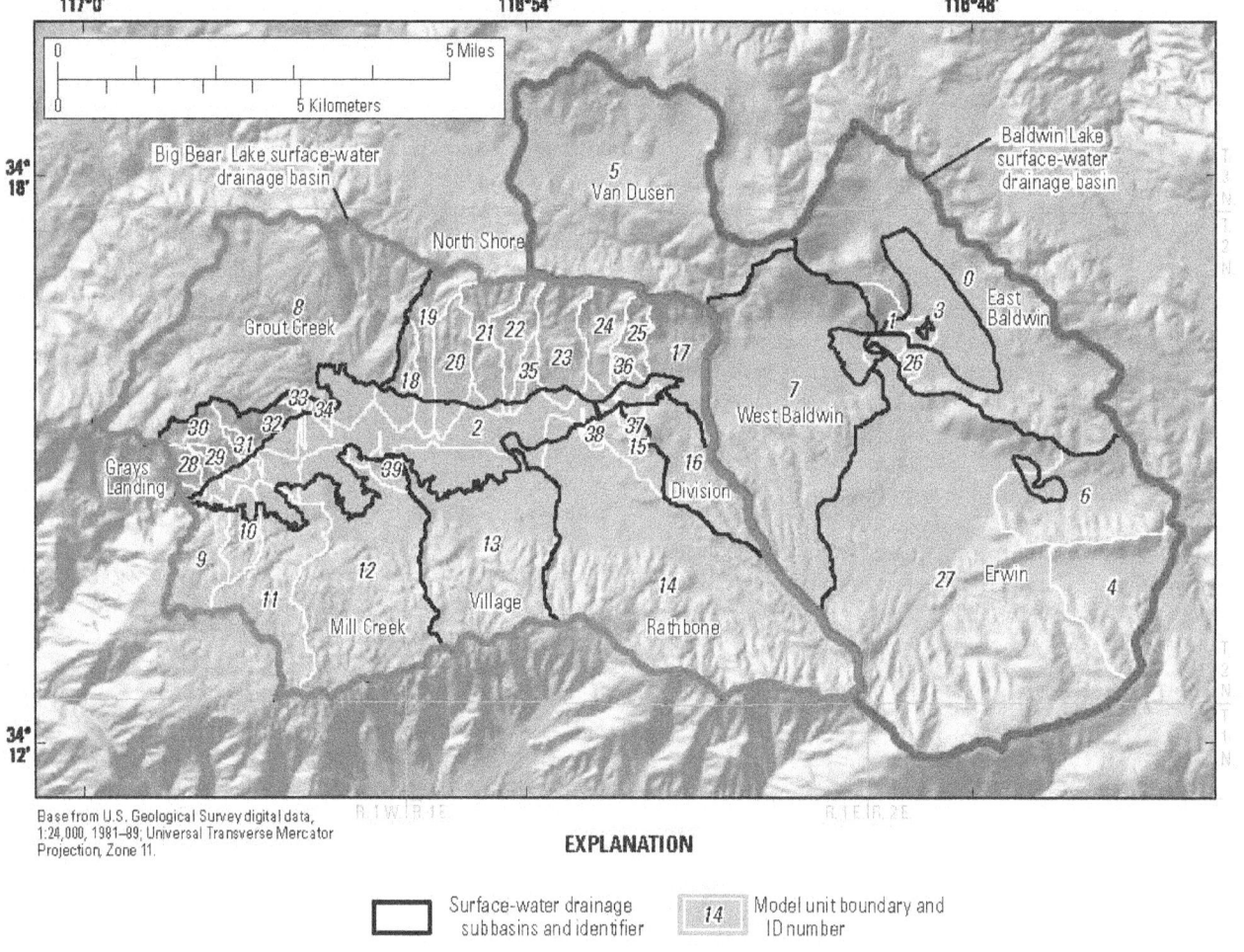

Base from U.S. Geological Survey digital data, 1:24,000, 1981–89; Universal Transverse Mercator Projection, Zone 11.

EXPLANATION

☐ Surface-water drainage subbasins and identifier

[14] Model unit boundary and ID number

Figure 19.   Spatial discretization for Big Bear and Baldwin Lakes surface-water drainage basins for INFILv3 model representing 12 subbasins, 40 model units, and the basin hydrography of Big Bear Valley, San Bernardino County, California.

Delineation of Surface-Water Drainage Basins

The INFILv3 model requires the discretization of the area being modeled into a 2-dimensional grid of equal-area (square) cells in the horizontal plane that are linked to create a surface-water routing network. For this study, the INFILv3 model grid was made equivalent to the 98.4-ft (30-m) DEM grid used to define the Big Bear Valley study area. The INFILv3 model domain contains a total of 46,122 acres (*table 4*), with about 92 percent of the area consisting of land and about 8 percent consisting of water (Big Bear Lake and Baldwin Lake).

The model area was delineated into 40 separate surface-water model units, with 31 model units defining the Big Bear Lake surface-water drainage basin and 9 model units defining the Baldwin Lake surface-water drainage basin (*fig. 19*). The 40 model units compose a linked network of tributary sub-drainages (bounded by thick lines in *fig. 19* with labeled names) upstream of Big Bear and Baldwin Lakes, and were used to define the 12 subbasins within the study area (8 sub-basins draining into Big Bear Lake and 4 draining into Baldwin Lake). Most of the model units in the Big Bear Lake surface-water drainage basin include both lake-area and land-area grid cells (*fig. 19*). Model unit 2 includes mostly lake-area cells and was used to collect all surface-water inflows into Big Bear Lake, including streamflow from the 30 upstream model units and runoff from the land areas within model unit 2. Model unit 3 includes only lake-area cells and was used to more efficiently accumulate all surface-water inflows into Baldwin Lake.

The distribution of DEM altitudes is similar for the two surface-water drainage basins, with an average altitude of 7,312 ft for Baldwin Lake surface-water drainage basin and 7,220 ft for Big Bear Lake surface-water drainage basin (*table 4*). The average slope of the DEM also is similar for the two basins—approximately 10 degrees for both basins (*table 4*). Model units with the highest average slopes (greater than 15 degrees) are located in the North Shore and Gray's Landing sub-drainages of the Big Bear Lake surface-water drainage basin.

The segmentation into 40 model units (including the two lakes) was done to improve the efficiency of the model operation, to accumulate all surface water discharging into the lake areas from the surrounding land areas, and to allow for a more direct comparison of model results with results for the drainage basins from previous studies (GeoScience Support Services, Inc., 2003) and the BCM results presented in this report.

Vertical Discretization And Layering

The root zone was discretized into vertical layers to account for differences in root density and root-zone water content with depth. Vertical discretization was defined for each grid cell by using two to five layers representing the soil component of the root zone. A sixth layer was used to represent consolidated bedrock for locations with thin soils or to designate the hydraulic conductivity of unconsolidated geologic units underlying the soil component for locations with thick soils. The INFILv3 model was modified for this study to include a seventh layer that represents a perched zone beneath the root zone (*fig. 18B*).

The number and thickness of layers defined for each grid cell depended on a combination of the estimated total root-zone thickness and the estimated soil thickness. Locations with thick soils were defined by using the areal extent of the mapped alluvial and unconsolidated deposits overlying the study area (*fig. 3*). On the basis of previous studies (Hevesi and others, 2003; Flint and Flint, 2007b), the thickness of the root zone was set to 26 ft for these locations. Drainage from the root zone was simulated as a function of the water content of layer 5 and the estimated hydraulic conductivity of the underlying material. For locations with thinner soils underlain by partially-consolidated or consolidated bedrock, the number of soil layers and the thickness of both the soil layers and the underlying bedrock layer (layer 6) were based on the estimated soil thickness and vegetation type (see Rewis and others, 2006, for a more detailed description). Layer 6 was used to represent the extension of the root zone into bedrock (roots extending into fractures and weathered zones). The thickness of the perched zone (layer 7) was dependent on rock type and was defined by dividing the storage capacity of the perched zone (2 ft) by the effective porosity estimated for layer 6. The resulting thickness of the perched zone varied from a minimum of 5.7 ft for unconsolidated deposits having effective porosities of 0.35 to a maximum of 40.0 ft for the consolidated rocks having an effective porosity of 0.05 (*table 5*).

Model Input and Data Requirements

Input to the INFILv3 model consists of five main input groups: (1) climate and meteorological data, (2) digital-map files and associated attribute tables used to define spatially-distributed parameters for individual grid cells, (3) model coefficients uniformly applied to all grid cells, (4) boundary conditions, and (5) initial conditions. Climate and meteorological data include daily input time series for precipitation and air temperature. Spatially-distributed parameters represent the physical characteristics of the drainages being modeled. Model coefficients include parameters used to model snowmelt and sublimation, to define stream-channel characteristics, and to define precipitation intensity using specified winter and summer storm durations. Boundary conditions are the daily surface-water inflows from model units upstream of the model unit being modeled (*table 4*). Initial conditions are the starting water contents of the root-zone layers, the perched zone, and the snowpack.

Table 4. INFILv3 model discretization and delineation of model area for Big Bear Valley, San Bernardino County, California.

[Abbreviations: DEM, digital elevation model; ft, feet; –, no data available]

Basin or subbasin name	Downstream subbasin or lake	Model unit	Downstream model unit or lake	Total area (acres)	Land area (acres)	Water area (acres)	Average altitude (ft)	Maximum altitude (ft)	Minimum altitude (ft)	Average slope (degrees)
				Discretization and layout of the INFILv3 model domain				DEM statistics		
Big Bear Lake surface-water drainage basin	Big Bear Lake	–	–	23,350	20,487	2,864	7,220	8,888	6,746	10.1
Big Bear Lake subbasin	Big Bear Lake	2	Big Bear Lake	1,571	418	1,153	6,770	7,804	6,747	2.3
Grout Creek subbasin	Big Bear Lake subbasin	8	2	4,512	4,298	214	7,315	8,379	6,748	12.2
Village subbasin	Big Bear Lake subbasin	13	2	2,456	2,050	406	7,121	8,173	6,746	8.2
Division subbasin	North Shore subbasin	16	17	797	789	8	6,989	7,416	6,747	7.1
Mill Creek subbasin	Big Bear Lake subbasin	–	–	4,510	4,049	461	7,230	7,949	6,746	8.1
		9	2	503	484	19	7,364	7,685	6,748	9.2
		10	2	258	200	58	7,043	7,536	6,748	11.3
		11	2	1,113	1,001	113	7,290	7,949	6,747	7.9
		12	2	2,529	2,280	250	7,215	7,935	6,746	7.8
		39	2	107	84	22	6,766	6,807	6,746	2.5
Rathbone subbasin	Big Bear Lake subbasin	–	–	4,863	4,789	74	7,419	8,888	6,748	12.1
		14	2	4,534	4,473	61	7,458	8,888	6,748	12.5
		15	2	242	234	8	6,891	7,097	6,748	7.0
		37	2	50	48	3	6,967	7,095	6,748	8.7
		38	2	37	34	2	6,789	6,849	6,748	3.1
North Shore subbasin	Big Bear Lake subbasin	–	–	3,783	3,404	380	7,134	8,209	6,747	11.7
		17	2	637	605	32	6,992	7,857	6,747	9.9
		18	2	279	161	118	6,869	7,385	6,748	6.9
		19	2	437	370	68	7,187	7,848	6,748	11.5
		20	2	432	354	78	6,986	7,618	6,748	8.6
		21	2	433	417	16	7,222	8,050	6,748	12.3
		22	2	313	306	8	7,257	8,209	6,748	12.8
		23	2	553	510	43	7,234	8,166	6,748	13.2
		24	2	369	367	2	7,344	7,974	6,748	16.9
		25	2	188	187	1	7,270	7,924	6,747	16.4
		35	2	101	89	12	6,921	7,176	6,748	6.1
		36	2	41	39	2	6,939	7,245	6,748	15.0

Table 4. INFILv3 model discretization and delineation of model area for Big Bear Valley, San Bernardino County, California.—Continued

[Abbreviations: DEM, digital elevation model; ft, feet; –, no data available]

Basin or subbasin name	Downstream subbasin or lake	Discretization and layout of the INFILv3 model domain					DEM statistics			
		Model unit	Downstream model unit or lake	Total area (acres)	Land area (acres)	Water area (acres)	Average altitude (ft)	Maximum altitude (ft)	Minimum altitude (ft)	Average slope (degrees)
Gray's Landing subbasin	Big Bear Lake subbasin	–	–	858	690	168	7,232	8,038	6,748	14.4
		28	2	103	101	1	7,382	8,015	6,748	21.3
		29	2	43	38	5	7,203	7,811	6,748	18.5
		30	2	264	237	28	7,473	8,038	6,748	14.3
		31	2	61	49	12	7,109	7,621	6,748	17.0
		32	2	305	213	92	7,096	7,932	6,748	12.7
		33	2	44	33	11	6,945	7,238	6,748	10.6
		34	2	37	19	19	6,795	6,946	6,748	6.0
Baldwin Lake surface-water drainage basin	Baldwin Lake	–	–	22,771	21,741	1,031	7,312	9,940	6,692	10.2
Van Dusen subbasin	West Baldwin subbasin	5	7	4,347	4,345	2	7,534	8,329	6,947	10.4
West Baldwin subbasin	East Baldwin subbasin	7	1	3,998	3,851	147	6,993	8,165	6,693	7.7
Erwin subbasin	West Baldwin subbasin	–	–	9,664	9,633	30	7,541	9,940	6,693	11.9
		4	6	1,113	1,104	9	7,567	8,779	7,161	12.2
		6	27	1,327	1,327	0	7,001	7,615	6,738	10.7
		27	7	7,223	7,202	21	7,637	9,940	6,693	12.1
East Baldwin subbasin	Baldwin Lake	–	–	4,763	3,912	851	6,909	8,213	6,692	8.5
		0	3	4,422	3,670	752	6,916	8,213	6,692	8.7
		1	3	232	173	58	6,878	7,374	6,693	8.5
		3	Baldwin Lake	2	0	2	6,693	6,694	6,693	0.3
		26	3	107	69	38	6,710	6,732	6,692	0.7
Big Bear study area	–	–	–	46,122	42,227	3,894	7,265	9,940	6,692	10.1

Table 5. Estimated effective porosity, saturated hydraulic conductivity, and thickness of the shallow groundwater zone used in the INFILv3 model of Big Bear Valley, San Bernardino County, California.

Age	Geology	Formation	Effective porosity	Layer 6 and 7 properties		Shallow groundwater-zone thickness (feet)
				Saturated hydraulic conductivity (feet/day)		
				Upper	Lower	
Cambrian	Carbonates	Bonanza King Formation	0.3	6.56E-01	1.31E-01	6.7
		Cararra Formation	0.05	6.56E-03	3.28E-04	40
		Wood Canyon Formation	0.05	6.56E-03	3.28E-04	40
	Quartzite		0.05	6.56E-04	3.28E-05	40
Cretaceous	Granites	Mixed diorite and gabbro	0.05	6.56E-03	3.28E-04	40
	Granites and gabbro		0.05	6.56E-02	1.31E-02	40
	Monzogranites		0.05	6.56E-04	3.28E-05	40
Jurassic	Granites	Bertha Peak pluton of Cameron (1981)	0.05	6.56E-03	3.28E-04	40
		Hornblende-biotite quartz diorite	0.05	6.56E-04	3.28E-05	40
		Gneissic granitoid rocks and gneiss	0.05	6.56E-04	3.28E-05	40
Mesozoic	Metamorphics		0.05	6.56E-04	3.28E-05	40
Proterozoic	Metamorphics		0.05	6.56E-03	1.31E-03	40
	Metamorphics		0.05	6.56E-04	3.28E-05	40
	Quartzite	Quartzite of Wildhorse Meadows	0.05	6.56E-04	3.28E-05	40
		Stirling Quartzite, carbonate and quartzite member	0.1	1.64E-01	3.28E-02	20
		Stirling Quartzite, carbonate-rich rocks	0.1	2.62E-02	1.31E-03	20
		Stirling Quartzite, quartzite member	0.05	6.56E-04	3.28E-05	40
Quaternary	Very old sediments	Very old alluvial fan deposits	0.2	3.28E+00	6.56E-01	10
	Old sediments	Old alluvial fan deposits	0.2	9.84E-01	1.97E-01	10
		Old alluvial fan deposits, Unit 3	0.2	1.64E+00	3.28E-01	10
		Old alluvial valley deposits	0.3	1.64E+00	3.28E-01	6.7
		Old landslide deposits	0.3	1.64E+00	3.28E-01	6.7
	Young sediments	Young alluvial fan deposits	0.3	6.56E+00	1.31E+00	6.7
		Young alluvial fan deposits, Unit 4	0.3	3.28E+00	6.56E-01	6.7
		Young alluvial valley deposits	0.3	3.28E+00	6.56E-01	6.7
		Young landslide deposits	0.3	3.28E+00	6.56E-01	6.7
		Young talus deposits	0.3	6.56E+00	1.31E+00	6.7
	Very young sediments	Very young alluvial fan deposits	0.1	1.64E+00	3.28E-01	20
		Very young alluvial fan deposits, Unit 1	0.35	1.64E+01	3.28E+00	5.7
		Very young alluvial valley deposits	0.35	1.64E+01	3.28E+00	5.7
		Very young colluvial deposits	0.35	3.28E+00	6.56E-01	5.7
		Very young lacustrine deposits	0.35	1.64E+00	3.28E-01	5.7
		Very young lacustrine deposits, Unit 2	0.1	1.64E+00	3.28E-01	20
		Very young talus deposits	0.35	1.64E+01	3.28E+00	5.7
		Very young wash deposits	0.35	6.56E+00	1.31E+00	5.7
Quaternary/tertiary	Breccia	Slide breccias	0.05	1.31E-02	6.56E-04	40
Tertiary	Granites	Alkalic granitic rocks, undifferentiated	0.05	6.56E-03	3.28E-04	40
	Sedimentary rocks	Sedimentary rocks of Holcomb Valley	0.05	6.56E-04	3.28E-05	40
		Sedimentary rocks of Poligue Canyon	0.1	3.28E-02	6.56E-03	20
		Sedimentary rocks south of Big Bear Lake	0.05	6.56E-02	3.28E-03	40
Triassic	Monzogranites		0.05	6.56E-04	3.28E-05	40

Climate Inputs

Daily climate data (precipitation and air temperature) are available from 144 climate stations in southern California for at least part of the period between January 1, 1927, and September 30, 2008. Data from these stations are collected and stored by different agencies, including the California Department of Water Resources' California Irrigation Management Information System stations (CIMIS), the National Climatic Data Center (NCDC), the National Interagency Fire Center's Remote Automated Weather Stations (RAWS), and San Bernardino County (SBC). The data from these stations were used to develop climate models for the study area.

The INFILv3 model estimates the daily precipitation and air-temperature values for each grid cell by spatial interpolation using a modified inverse-distance-squared interpolation algorithm (Hevesi and others, 2003). For this study, the algorithm to estimate precipitation was modified to allow for the estimation of monthly PRISM data (Daly and others, 1994, 2004) using the daily records from the climate stations. The monthly PRISM data consist of average monthly precipitation maps available for the nation on an approximate 2,625-ft (800-m) grid spacing for the 30-year period 1971–2000 (Daly and others, 1994, 2004). The data were downscaled to each INFILv3 98.4-ft (30-m) grid cell for Big Bear Valley. The monthly PRISM estimates incorporate multiple variables in order to account for complex orographic effects on precipitation, such as rain shadows and adiabatic cooling. The modification to INFILv3 to incorporate the PRISM data was considered to be an improvement relative to the simple precipitation-altitude regression models that have been used previously (Hevesi and others, 2003).

For calibration purposes, an initial spatial-interpolation model for daily precipitation and air temperature, herein referred to as the "preliminary climate model," was developed using records from 35 selected NCDC climate stations (*fig. 20A, table 6*) and the monthly PRISM data. The dataset, consisting of the 35 selected records, is referred to in the study as the "preliminary climate input." The preliminary climate input used only the NCDC records because NCDC incorporates a rigorous quality control review process of climate data archived on its database (EarthInfo, Inc., 2006). The selection of the 35 stations was based on proximity to the study site and adequacy of record (only stations having 3 or more years of record were included in the network). Evaluation of the non-NCDC daily climate records from stations in the study area indicated multiple data gaps and inconsistencies in the timing and magnitude of daily precipitation, even when comparing records for stations in close proximity. The inconsistencies in the records were attributed to several possible factors, including differences in the frequency and timing of data collection and difficulties in measuring precipitation occurring as snow. Monthly precipitation amounts, however, were more consistent, even for stations in close proximity. A limitation of utilizing only the NCDC data was that several non-NCDC stations located in the Baldwin Lake surface-water drainage basin were omitted from the preliminary climate input.

Climate data were compiled for this study for water years 1928–2005. A minimum of two stations had data for any given date in the simulation. The period having the greatest number of stations with data was approximately 1970–99. On the basis of an analysis of the number of stations having data for a given date, the period from October 1, 1949, to September 30, 2005 (water-years 1950–2005), was determined to be the most appropriate for application of the preliminary climate inputs for simulating water balance (including recharge) in the Big Bear Valley study area.

The spatial distribution of average annual precipitation estimated for water years 1950–2005 by using INFILv3 with the preliminary climate input (*fig. 20B*) is similar to that generated by PRISM (*fig. 2*). The INFILv3 estimated precipitation ranges from about 34 inches per year (in/yr) for the summit areas on the southeast and southwest edges of the study area to about 18 to 20 in/yr along the northeastern edge of the study area (*fig. 20B*). The spatial distribution represents the combined effects of precipitation sources (most storms track from the west and southwest), adiabatic cooling as storms are forced over higher altitudes, and rain shadow effects on the leeward side of mountains (the northeastern boundary bordering Baldwin Lake playa). The spatial distribution is not identical to that of PRISM because the daily INFILv3 simulation honors the available measured precipitation records at the NCDC climate stations, whereas PRISM is an average for the period 1971–2000. The effect of the PRISM maps on the INFILv3 spatial interpolation increases with increasing distance from the nearest precipitation stations.

Spatially-interpolated average annual precipitation, estimated for water years 1950–2005 by using the preliminary climate model as input to INFILv3, closely matched the measured precipitation from the two NCDC stations in the Big Bear surface-water drainage basin; however, the estimates are higher than the measured values at many non-NCDC stations in the valley by as much as 0.1 in/day (*fig. 20B*). Analysis of the uncertainty in the preliminary climate model and inputs, including the discrepancy between estimated and observed precipitation for non-NCDC stations, is presented in later sections of this report.

Daily air temperature did not require conditioning to PRISM maps for maximum and minimum air temperature because adiabatic cooling is the primary orographic process affecting air temperature in the Big Bear area, which causes air temperature to be strongly correlated with altitude. This strong correlation allows for the development of regression models for average monthly maximum and minimum air temperature by using altitude as the independent variable. The monthly regression models were applied instead of the PRISM maps to define the average monthly maximum and minimum air temperatures needed as part of the preliminary climate input. The average monthly regression models were used to condition the linear weighting factors used in the inverse-distance-squared interpolation of the daily air-temperature records (both maximum and minimum daily air temperature) over the model area (Rewis and others, 2006; Hevesi and others, 2003).

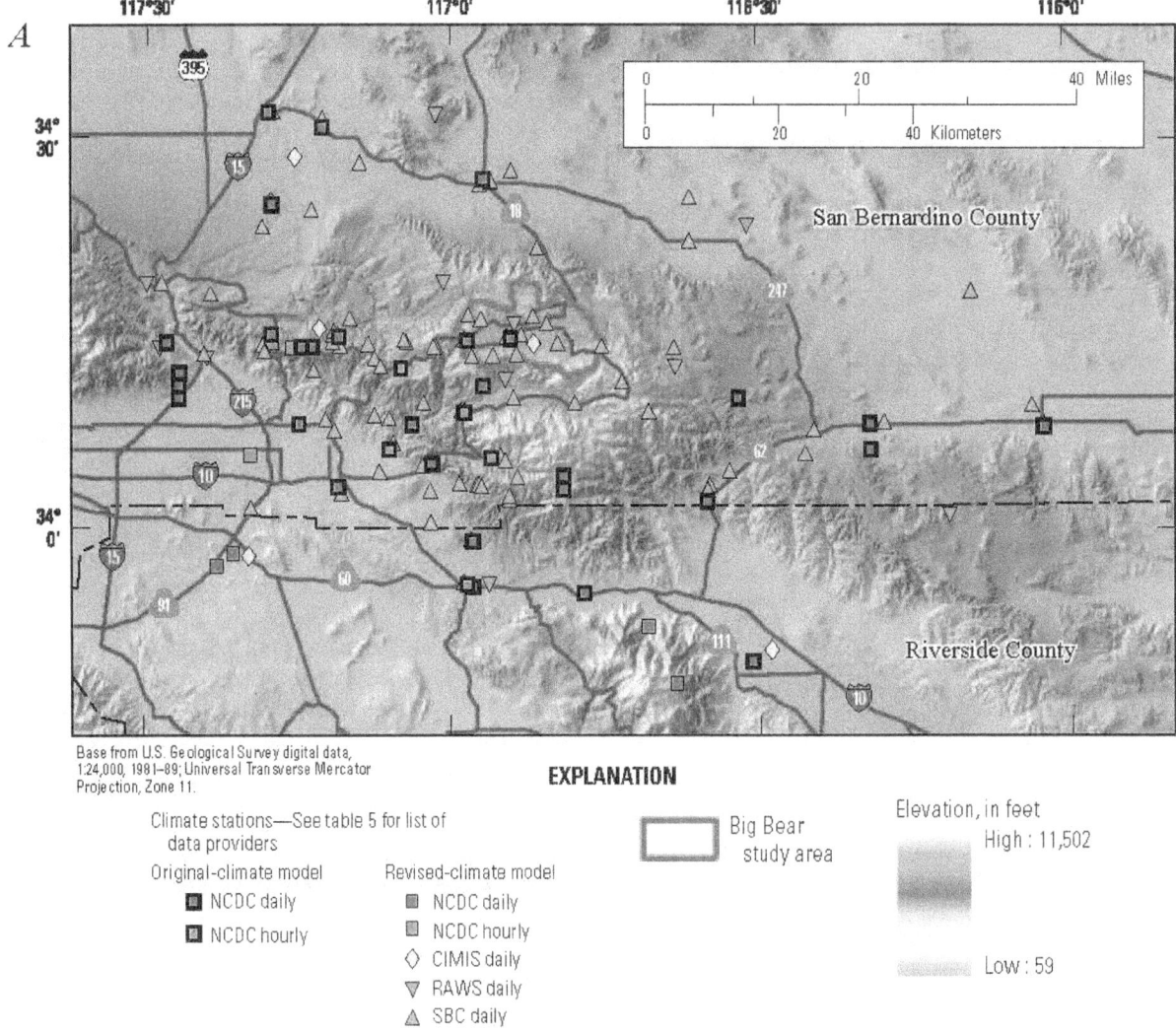

Figure 20. (A) Meteorological stations used to develop climate models and (B) INFILv3 average annual precipitation for water years 1950–2005 simulated by using the preliminary climate model for Big Bear Valley, San Bernardino County, California.

Average annual air temperature estimated for water years 1950–2005 by using INFILv3 and the preliminary climate input were characterized by a spatial distribution closely matching topography (fig. 21). Minimum average annual air temperatures less than 38°F were estimated for the high altitude locations in the southeastern portions of the study area. Maximum average air temperatures of 46 to 47°F were estimated for the low altitude areas surrounding and including Big Bear and Baldwin Lakes. The spatial distribution of estimated air temperature is a critical factor affecting the estimation of precipitation occurring as snow, snowmelt, sublimation, and potential evapotranspiration.

Model Parameterization

Digital map files and standard GIS methods were used to develop most of the input parameters required for the INFILv3 model to represent the physical characteristics of the drainage basins. Maps used as input included the grid-formatted DEM of the study area as well as the vector-formatted maps of soil type, surface geology (fig. 3), and vegetation type. The digital map files were used to define the spatial distribution of drainage basin parameters for INFILv3, including (1) topographic parameters, (2) vegetation and root-zone parameters, (3) soil parameters, and (4) rock parameters. Attribute tables were used to define the properties corresponding to the parameters. Geologic maps from three sources were used in this study. The geologic map of the Fawnskin 7.5′ quadrangle (Miller and others, 2001) was used for parts of the south shore of Big Bear Lake and the area north of Big Bear Lake. The adjacent geologic map of the Big Bear City 7.5′ quadrangle (Miller, 2004) was used for Baldwin Lake and the northern part of Bear Valley. For the area south of latitude 34°15′N, an unpublished compilation of the geology prepared for the U.S. Forest Service at a scale of 1:100,000 (D.M. Morton, U.S. Geological Survey, written commun., 2005) was used.

B

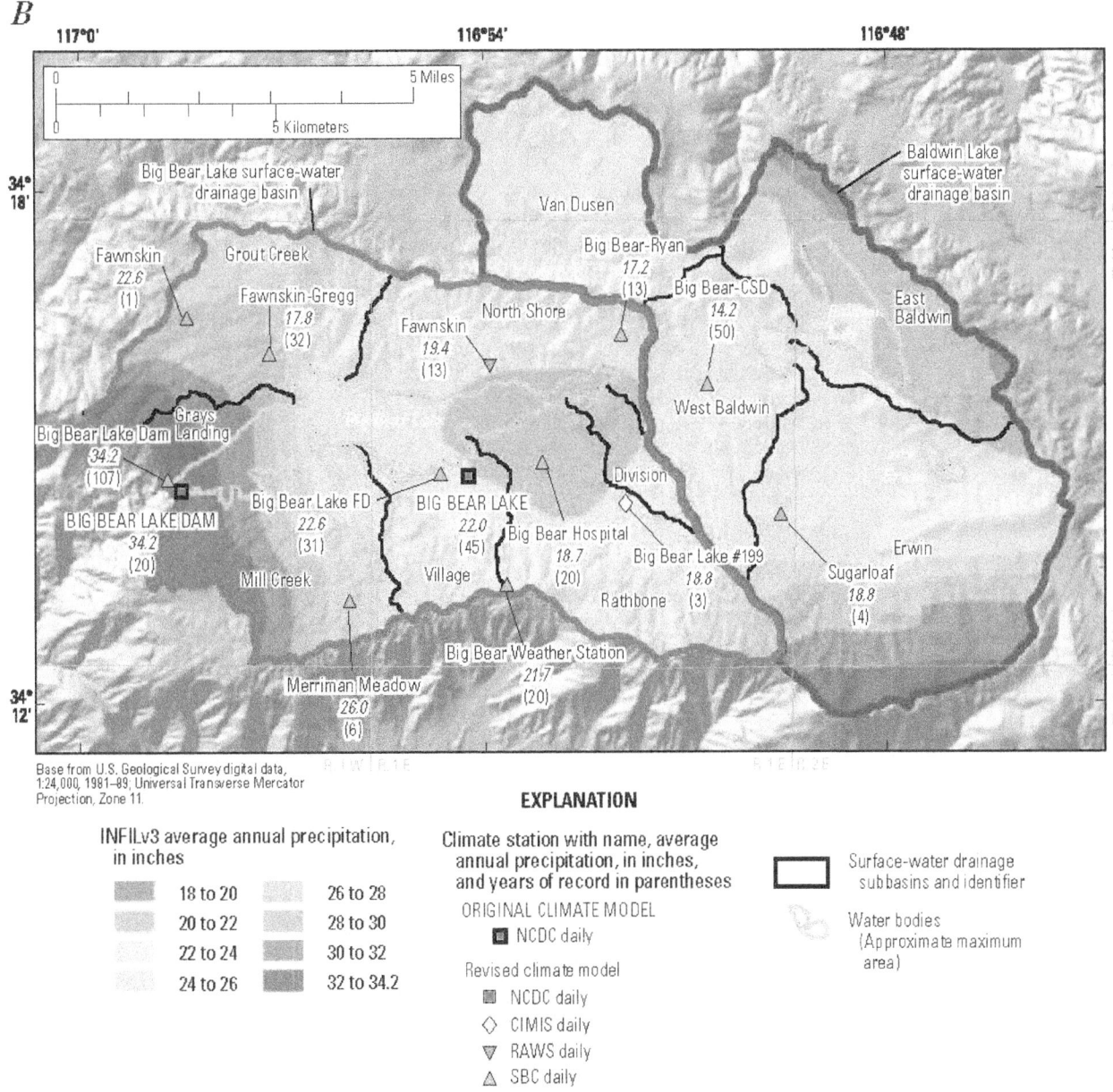

Figure 20. Continued

Topographic parameters are used in INFILv3 to simulate potential evapotranspiration, to estimate spatially-distributed air temperature, and to route runoff as overland flow and streamflow. The DEM of the study area was used to define the topographic parameters for each model cell (Hevesi and others, 2003), including altitude, aspect, slope, the skyview parameter (used to simulate incoming solar radiation), a set of 36 blocking ridge angles (used to simulate the effects of shading on potential evapotranspiration in rugged areas), and the runoff-routing parameters (location of upstream cell, location of downstream cell, and number of upstream cells). The flow-routing parameters were calculated by using ARC-Hydro and were based on a routing algorithm that represents convergent flow only (a given cell can route to only one downstream cell, but can receive inflows from multiple upstream cells).

Table 6. Meteorological stations used in development of climate models for daily INFILv3 simulations for Big Bear Valley, San Bernardino County, California.

[NCDC stations used in model calibration. Supplemental stations added for climate model revisions. Altitude in National Geodetic Vertical Datum of 1929. **Abbreviations**: CIMIS, California Irrigation Management and Information System; NCDC, National Climate Data Center; RAWS, Remote Automated Weather Station; SBC, San Bernardino County; UTM, Universal Transverse Mercator coordinate system]

Station name	Data source/ time interval	Station number	UTM Easting location (meters)	UTM Northing location (meters)	Altitude (feet)	Years of precipitation record	Years of air temperature record
NCDC stations used in preliminary and revised climate models							
Apple Valley	NCDC/daily	40244	480,110	3,819,471	2,935	27	0
Beaumont Pumping Plant	NCDC/daily	40607	503,076	3,760,305	3,051	27	17
Beaumont #2	NCDC/daily	40609	503,059	3,753,874	2,600	53	52
Bennett Ranch	NCDC/daily	40678	458,525	3,780,731	1,850	5	0
Big Bear Lake	NCDC/daily	40741	508,849	3,789,237	6,760	45	45
Big Bear Lake Dam	NCDC/daily	40742	502,357	3,788,923	6,815	20	9
Cabazon	NCDC/daily	41250	520,031	3,752,941	1,801	23	0
Crestline	NCDC/daily	42162	472,377	3,789,917	4,872	3	0
Fontana 5 N	NCDC/daily	43118	458,533	3,782,572	1,972	31	1
Hesperia	NCDC/daily	43935	472,432	3,808,401	3,202	18	0
Joshua Tree	NCDC/daily	44405	563,003	3,777,147	2,723	14	0
Joshua Tree 3 S	NCDC/daily	44407	563,028	3,773,455	3,491	3	0
Kee Ranch	NCDC/daily	44467	543,015	3,780,738	4,334	27	0
Lake Arrowhead	NCDC/daily	44671	482,661	3,789,526	5,205	57	57
Lucerne Valley	NCDC/daily	45182	504,593	3,812,054	2,963	23	23
Lytle Creek PH	NCDC/daily	45215	458,541	3,784,424	2,251	19	0
Lytle Creek R S	NCDC/daily	45218	456,644	3,788,679	2,730	52	0
Mill Creek 2	NCDC/daily	45629	496,928	3,771,393	2,943	19	19
Morongo Valley	NCDC/daily	45863	538,466	3,765,927	2,562	18	0
Palm Springs	NCDC/daily	46635	545,370	3,743,138	425	77	77
Raywood Flats	NCDC/daily	47279	516,917	3,767,715	7,073	11	0
Redlands	NCDC/daily	47306	482,520	3,768,027	1,318	78	78
San Bernardino F S 226	NCDC/daily	47723	476,590	3,777,087	1,140	75	75
Santa Ana River PH 1	NCDC/daily	47894	493,850	3,776,938	2,772	19	10
Seven Oaks	NCDC/daily	48105	504,607	3,782,481	5,082	4	3
South Fork Cabin	NCDC/daily	48390	516,914	3,769,567	7,126	4	0
Squirrel Inn 1	NCDC/daily	48476	476,976	3,788,053	5,243	11	0
Squirrel Inn 2	NCDC/daily	48479	478,514	3,788,049	5,682	23	23
Twentynine Palms	NCDC/daily	49099	588,808	3,776,779	1,975	56	57
Victorville Pump Plant	NCDC/daily	49325	471,938	3,821,521	2,858	58	58
Beaumont	NCDC/hourly	100606	502,311	3,754,306	2,613	46	0
Camp Angelus	NCDC/hourly	101369	501,816	3,778,699	5,770	53	0
Mill Creek Intake	NCDC/hourly	105632	505,867	3,772,292	4,945	49	0
Running Springs 1 E	NCDC/hourly	107600	492,041	3,785,078	5,965	53	0
Santa Ana River PH 3	NCDC/hourly	107891	490,214	3,773,438	1,984	40	0
Supplemental stations used in revised climate model							
Mount San Jacinto Wsp	NCDC/daily	45978	533,943	3,740,041	8,425	10	9
Riverside Fire Sta 3	NCDC/daily	47470	464,139	3,756,802	840	78	78
Riverside Citrus Exp St	NCDC/daily	47473	466,613	3,758,545	986	54	51
Snow Creek Upper	NCDC/daily	48317	529,623	3,748,065	1,940	55	0
Crestline Lake Gregory	NCDC/hourly	102163	475,438	3,788,057	4,534	12	0
Crestline	NCDC/hourly	102164	472,475	3,788,719	4,870	33	0
Lytle CK Fthill Blvd	NCDC/hourly	105212	469,126	3,772,740	1,160	43	0
Beaumont	RAWS/daily	600002	505,571	3,754,458	2,680	5	5
Big Pine Flat	RAWS/daily	600003	498,799	3,797,576	6,861	7	7
Burns Canyon	RAWS/daily	600004	533,726	3,785,532	6,000	17	17
Converse	RAWS/daily	600010	508,011	3,783,689	5,618	11	11
Devore	RAWS/daily	600015	462,747	3,786,747	2,057	16	17
Fawnskin	RAWS/daily	600018	509,308	3,791,667	6,900	13	13

Table 6. Meteorological stations used in development of climate models for daily INFILv3 simulations for Big Bear Valley, San Bernardino County, California.—Continued

[NCDC stations used in model calibration. Supplemental stations added for climate model revisions. Altitude in National Geodetic Vertical Datum of 1929. **Abbreviations**: CIMIS, California Irrigation Management and Information System; NCDC, National Climate Data Center; RAWS, Remote Automated Weather Station; SBC, San Bernardino County; UTM, Universal Transverse Mercator coordinate system]

Station name	Data source/ time interval	Station number	UTM Easting location (meters)	UTM Northing location (meters)	Altitude (feet)	Years of precipitation record	Years of air temperature record
Supplemental stations used in revised climate model—Continued							
Granite Mountain	RAWS/daily	600020	497,629	3,821,540	4,720	14	14
LostHorse	RAWS/daily	600023	574,992	3,764,425	4,200	17	17
LytleCreek	RAWS/daily	600024	455,769	3,788,194	2,792	7	7
MeansLake	RAWS/daily	600025	544,404	3,805,567	2,900	13	13
MillCreek	RAWS/daily	600026	496,797	3,771,427	2,950	11	11
MormonRock	RAWS/daily	600028	453,820	3,797,475	3,300	9	9
U.C. Riverside #44	CIMIS/daily	700044	468,999	3,758,326	1,020	23	23
Victorville #117	CIMIS/daily	700117	476,023	3,815,226	2,890	15	15
Cathedral City #118	CIMIS/daily	700118	548,212	3,744,815	392	12	13
Lake Arrowhead #192	CIMIS/daily	700192	479,847	3,790,760	5,148	5	4
Big Bear Lake #199	CIMIS/daily	700199	512,407	3,788,560	6,910	3	3
Camp Angelus	SBC/daily	800832	501,851	3,778,784	5,780	8	0
Devore-Wilmuth	SBC/daily	802011	463,301	3,795,390	2,500	66	0
Cajon Junction	SBC/daily	802016	455,972	3,796,863	3,118	65	0
Devore C.D.F.	SBC/daily	802118	462,540	3,786,850	2,080	57	8
Highgrove Steam Plant	SBC/daily	802222	469,329	3,764,788	945	48	0
Manzanita Flat	SBC/daily	802833	495,725	3,779,871	3,920	14	0
Panorama Point	SBC/daily	802840	471,405	3,787,337	3,775	70	0
Strawberry Creek	SBC/daily	802881	478,933	3,784,483	2,907	6	0
Yucaipa Ridge	SBC/daily	802900	509,871	3,769,131	9,020	6	4
Oak Creek Canyon	SBC/daily	802994	490,274	3,777,546	3,676	5	0
Oak Glen-Bise	SBC/daily	803014	503,897	3,768,152	4,680	37	0
Oak Glen	SBC/daily	803015	504,364	3,767,906	4,680	63	0
Camp Angelus	SBC/daily	803053	501,856	3,778,800	5,770	56	0
Mill Creek Ranger Station	SBC/daily	803077	495,591	3,770,927	2,980	44	4
Oak Glen-Sample	SBC/daily	803121	501,223	3,768,282	3,695	24	9
Oak Glen-Wagoner	SBC/daily	803122	504,082	3,768,263	4,040	43	0
Yucaipa C.D.F.	SBC/daily	831291	496,716	3,767,316	2,660	29	10
Santa Ana P.H. #3	SBC/daily	803162	490,930	3,774,096	1,950	84	4
Patton-George	SBC/daily	831701	480,658	3,777,484	1,375	32	0
Forest Falls	SBC/daily	803173	507,313	3,771,990	5,300	12	0
Forest Falls	SBC/daily	831731	507,309	3,771,955	5,300	24	0
Heart Bar Federal Park	SBC/daily	803259	518,665	3,779,809	6,688	43	0
Camp Angelus	SBC/daily	803260	501,400	3,778,798	5,780	39	17
Fallsvale	SBC/daily	803283	508,058	3,771,683	5,990	38	0
Highland-Dundee	SBC/daily	803315	481,946	3,775,929	1,205	16	0
Redlands-Bottenberg	SBC/daily	803329	483,027	3,766,906	1,465	26	11
Mentone C.D.F.	SBC/daily	803337	488,803	3,769,965	1,765	57	7
Oak Glen Conservation C	SBC/daily	803345	508,643	3,765,862	5,450	28	6
Oak Glen Conservation C	SBC/daily	803346	508,624	3,766,190	5,450	34	0
Calimesa-Raisner	SBC/daily	803386	496,883	3,762,639	2,620	18	13
Hesperia Pump Plant #22	SBC/daily	804002	471,079	3,804,948	3,380	14	0
Apple Valley-Rock Springs	SBC/daily	804003	478,617	3,807,394	2,890	14	0
Victorville Pump Plant #4	SBC/daily	804096	472,754	3,821,299	2,945	9	0
Apple Valley	SBC/daily	804136	480,196	3,820,329	2,930	4	0
Hesperia C.D.F.	SBC/daily	804195	472,357	3,808,886	3,175	40	4
Apple Valley County Yard	SBC/daily	804325	485,806	3,814,083	3,080	34	0
Baine Ranch Baker Hill	SBC/daily	804733	533,456	3,787,722	2,700	4	0
Lake Arrowhead FS #1	SBC/daily	805140	482,512	3,789,613	5,205	79	16

Table 6. Meteorological stations used in development of climate models for daily INFILv3 simulations for Big Bear Valley, San Bernardino County, California.—Continued

[NCDC stations used in model calibration. Supplemental stations added for climate model revisions. Altitude in National Geodetic Vertical Datum of 1929. **Abbreviations**: CIMIS, California Irrigation Management and Information System; NCDC, National Climate Data Center; RAWS, Remote Automated Weather Station; SBC, San Bernardino County; UTM, Universal Transverse Mercator coordinate system]

Station name	Data source/ time interval	Station number	UTM Easting location (meters)	UTM Northing location (meters)	Altitude (feet)	Years of precipitation record	Years of air temperature record
				Supplemental stations used in revised climate model—Continued			
Lake Arrowhead-Asher	SBC/daily	805209	481,971	3,789,568	5,360	34	14
Lake View Point	SBC/daily	805263	497,444	3,787,884	7,105	3	0
Green Valley Lake	SBC/daily	805264	492,655	3,788,662	6,880	23	0
Lake Arrowhead FS #2	SBC/daily	805281	484,369	3,791,832	5,200	35	17
Heaps Peak	SBC/daily	805339	487,032	3,788,157	6,421	32	0
Glen Crest	SBC/daily	858021	471,287	3,788,038	5,080	17	5
Twin Peaks-Crabtree	SBC/daily	805818	482,900	3,787,960	5,690	22	2
Crest Park Lutheran Church	SBC/daily	805819	481,692	3,788,459	5,525	18	9
Running Springs-Nob Hill	SBC/daily	805820	487,974	3,786,161	6,520	14	7
Luring Pines	SBC/daily	805824	488,967	3,785,007	6,240	11	5
Kuffel Canyon	SBC/daily	805834	482,006	3,789,905	5,450	10	7
Running Springs West	SBC/daily	805836	488,020	3,778,059	6,180	9	7
Lucerne Valley Cemetery	SBC/daily	806001	504,196	3,811,116	2,946	17	0
Yucca Valley-Alta Loma	SBC/daily	806006	553,212	3,772,579	3,740	13	0
Big Bear Lake Dam	SBC/daily	906032	502,284	3,788,909	6,815	107	18
Twentynine Palms	SBC/daily	860481	588,823	3,776,886	1,975	72	0
Lucerne Valley	SBC/daily	860571	505,708	3,811,341	2,957	20	0
Camp Oakes	SBC/daily	806070	522,671	3,787,902	7,450	2	0
Big Bear Lake FD	SBC/daily	806090	508,221	3,789,261	6,745	31	0
Big Bear CSD	SBC/daily	860911	514,359	3,791,145	6,800	50	10
Joshua Tree	SBC/daily	861341	565,123	3,777,132	2,760	33	0
Morongo Valley	SBC/daily	806135	538,789	3,767,784	2,570	32	0
Cushenberry Springs	SBC/daily	806224	512,895	3,801,866	4,250	39	7
Johnson Valley-WCS	SBC/daily	806255	535,644	3,809,103	2,794	37	6
Lucerne Valley FD	SBC/daily	806324	505,700	3,811,287	2,957	14	0
Big Bear-Ryan	SBC/daily	806330	512,400	3,792,213	7,000	13	0
Fawnskin-Gregg	SBC/daily	806334	504,344	3,791,769	6,820	32	18
Morongo Valley Trailer Park	SBC/daily	806354	541,693	3,770,017	2,765	10	4
Big Bear Hospital	SBC/daily	806363	510,517	3,789,445	6,800	20	11
Lucerne Valley Midway Park	SBC/daily	806372	508,932	3,812,946	2,910	10	4
Twentynine Palms U.S.M.C.	SBC/daily	806402	578,054	3,795,751	2,004	24	6
Big Bear Weather Station	SBC/daily	807000	509,830	3,786,687	8,188	3	2
Morongo Ridge	SBC/daily	807017	529,628	3,778,588	8,070	2	0
Beaumont	SBC/daily	807029	501,780	3,754,289	2,613	65	0
Beaumont Pumping Plant	SBC/daily	807030	502,887	3,760,277	3,045	29	0
Camp Tahquitz	SBC/daily	807715	509,288	3,780,634	6,560	5	0
Grace Valley	SBC/daily	807718	525,621	3,782,719	8,120	4	0
Sugarloaf	SBC/daily	807719	516,048	3,788,341	7,200	4	0
Merriman Meadow	SBC/daily	807720	506,194	3,786,555	7,530	6	0
Lake View Point	SBC/daily	807721	497,536	3,787,884	6,720	5	0
Bluff Lake	SBC/daily	807723	502,937	3,786,501	7,600	3	0
Yucca Valley C.D.F.	SBC/daily	809002	554,486	3,776,032	3,420	51	7
Twentynine Palms CY	SBC/daily	809004	587,092	3,779,440	1,895	48	11
Morongo Valley post office	SBC/daily	809010	538,580	3,767,232	2,580	17	0
Johnson Valley-MWA	SBC/daily	809012	535,599	3,802,893	2,950	11	0
Fawnskin	SBC/daily	809022	502,513	3,792,389	7,200	1	0
Green Valley FD	SBC/daily	809024	492,866	3,788,547	6,900	34	0

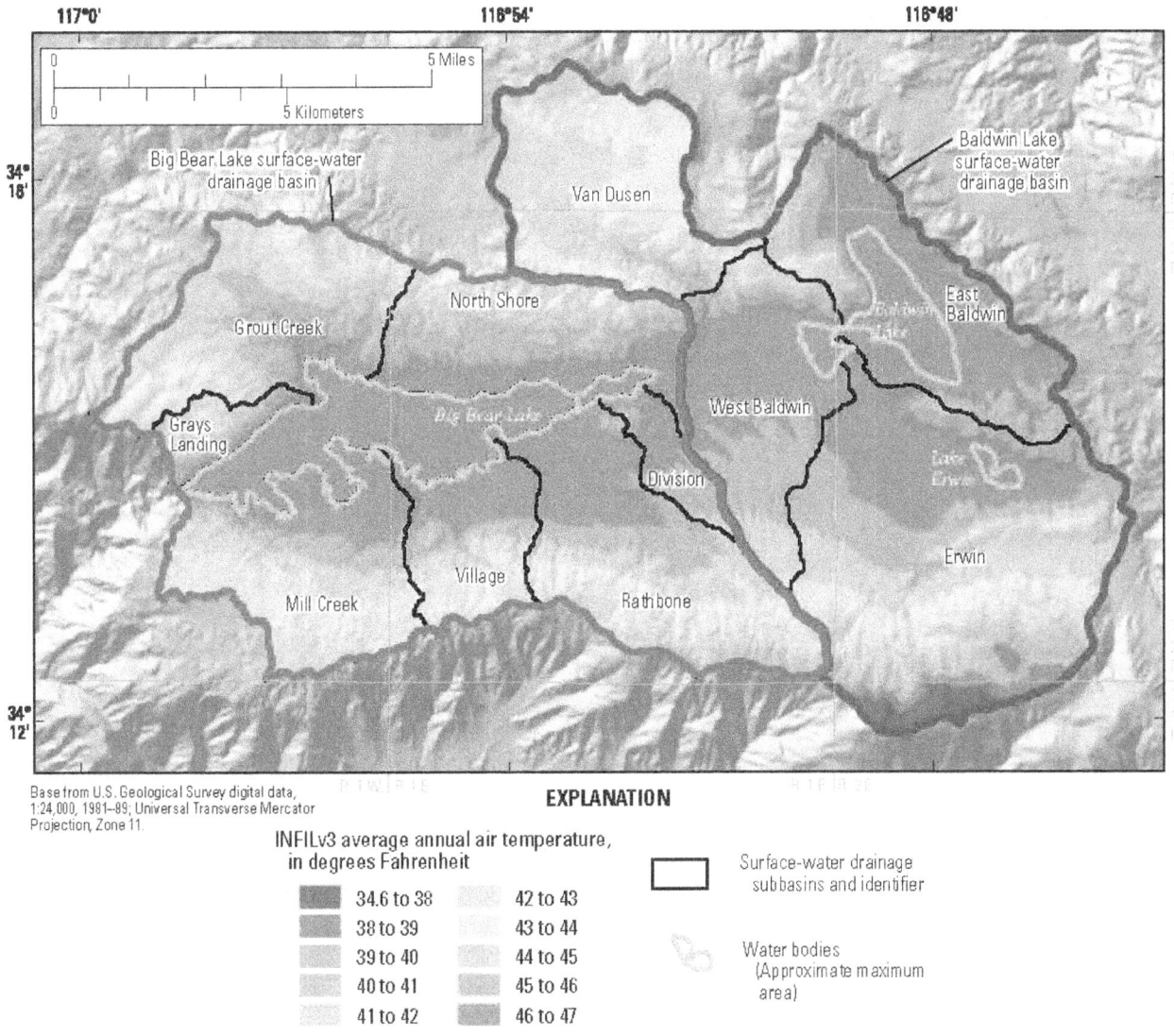

Figure 21. INFILv3 average air temperature for water years 1950–2005 simulated by using the preliminary climate model for Big Bear Valley, San Bernardino County, California.

Vegetation density estimates were based on a forest canopy map from the USGS seamless website (*http://seamless.usgs.gov/Website/Seamless/products/nlcd01.asp#description*). Root-zone thickness is defined using vegetation parameters, unless soil-zone thickness limits the thickness of the root zone to less than or equal to that of the soil zone. If soil-zone thickness is limiting and vegetation is primarily a tree or shrub type, bedrock can be included in the root-zone thickness, allowing tree or shrub roots to penetrate into bedrock (Hevesi and others, 2003; Rewis and others, 2006). A total of 25 different vegetation and land-use types were identified for the Big Bear Valley model area (*fig. 22* and *table 7*).

Soil parameters were estimated for each model cell by using the State Soil Geographic Database (STATSGO) digital map and associated attribute tables compiled by the U.S. Department of Agriculture (1994). The STATSGO data were the same soils data that were used for the BCM; however, the INFILv3 inputs incorporated a more detailed geology map (*fig. 3*) to define the location of alluvial deposits used to estimate soils greater than 6 feet. The soil parameters included physical and hydraulic properties calculated by using the STATSGO data (Hevesi and others, 2003): soil depth, porosity, field capacity, wilting-point water content, Brooks-Corey parameters for drainage (air-entry potential and drainage coefficient), and saturated hydraulic conductivity (K_{sat}; *table 8*). Soil parameters used in the INFILv3 model are average values for each STATSGO map-unit-identifier (MUID). The Big Bear Valley area contains five MUIDs (*fig. 23* and *table 8*).

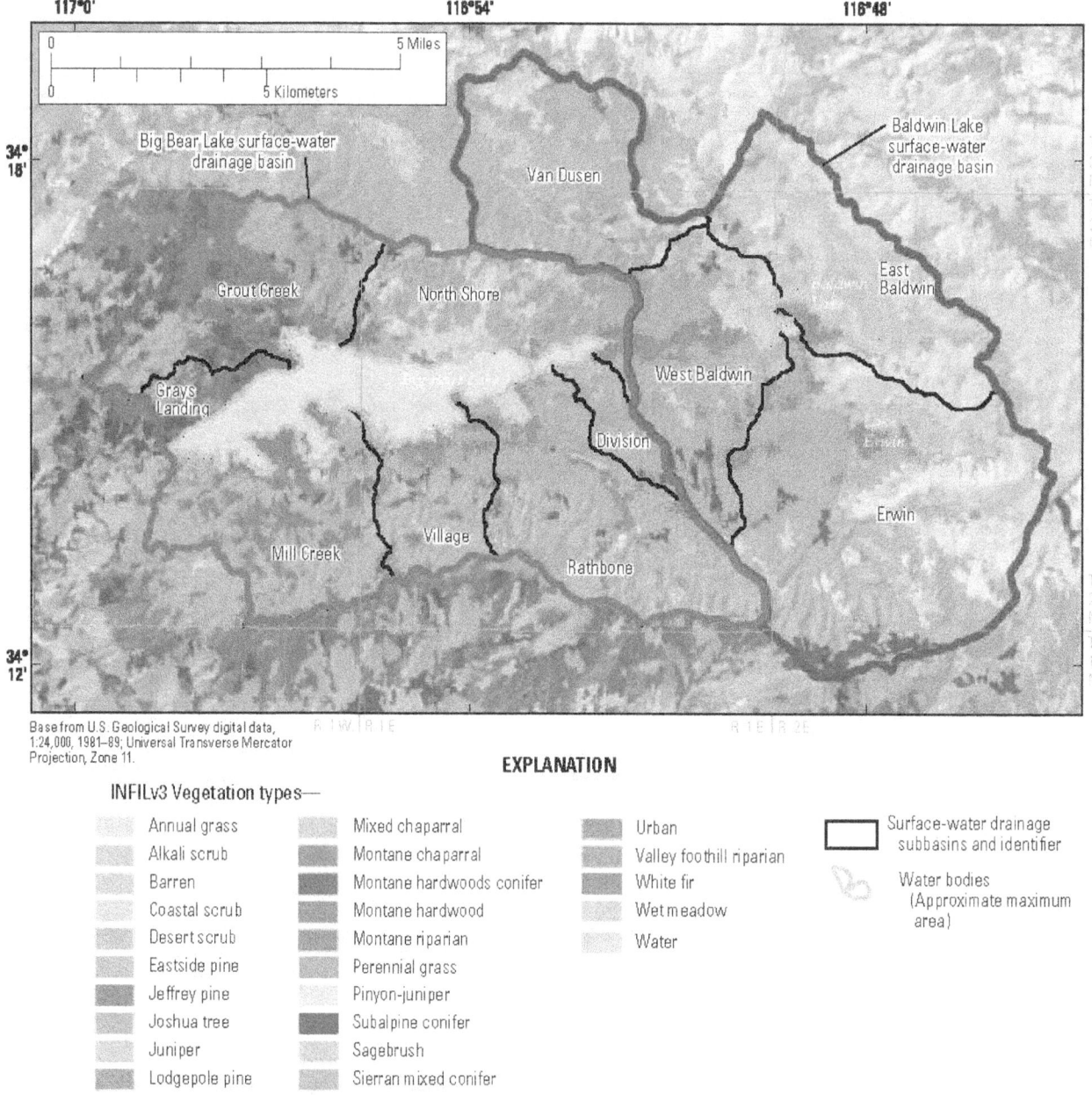

Figure 22. Vegetation and land-use types used for estimating vegetation density for the INFILv3 model for Big Bear Valley, San Bernardino County, California.

Parameters representing the properties of geologic units underlying the soil zone were estimated for each of the geologic units delineated on *figure 3*. The parameters included effective porosity, upper and lower saturated hydraulic conductivity, and thickness of the shallow groundwater zone (*table 5*). Estimates of effective porosity and upper and lower saturated hydraulic conductivity were based on a general knowledge of the characteristics of the different geologic units. For example, unconsolidated deposits were assumed to have a higher effective porosity and saturated hydraulic conductivity compared to consolidated rocks, and sedimentary rocks were assumed to have higher saturated hydraulic conductivity relative to igneous and metamorphic rocks. The thickness of the shallow groundwater zone was defined by assuming a storage capacity of 24 inches and then using the estimated effective porosity to calculate a thickness.

Table 7. Vegetation and root-zone parameters used in the INFILv3 model for Big Bear Valley, San Bernardino County, California.

Vegetation type	Estimated vegetation cover (percent)	Root-zone layer density (percent)					Maximum root-zone thickness (feet)					Maximum bedrock layer thickness (feet)
		Layer 1	Layer 2	Layer 3	Layer 4	Layer 5	Layer 1	Layer 2	Layer 3	Layer 4	Layer 5	Layer 6
Alkali scrub	0.5	55	33	22	11	11	0.3	1.3	4.9	6.6	13.1	6.6
Annual grass	1.3	28	17	11	6	6	0.3	1.3	4.9	6.6	13.1	6.6
Barren	3	8	7	6	6	6	0.3	1.3	4.9	6.6	13.1	6.6
Coastal scrub	17	70	42	28	14	14	0.3	1.3	4.9	6.6	13.1	6.6
Desert scrub	0.3	31	19	12	6	6	0.3	1.3	4.9	6.6	13.1	6.6
Eastside pine	29.5	61	55	49	43	43	0.3	1.3	4.9	6.6	13.1	13.1
Jeffrey pine	30.2	62	56	50	44	44	0.3	1.3	4.9	6.6	13.1	13.1
Joshua tree	2.6	23	16	11	7	7	0.3	1.3	4.9	6.6	13.1	9.8
Juniper	7.4	61	43	31	18	18	0.3	1.3	4.9	6.6	13.1	9.8
Lodgepole pine	46.5	95	85	76	66	66	0.3	1.3	4.9	6.6	13.1	13.1
Mixed chaparral	25.3	53	47	42	37	37	0.3	1.3	4.9	6.6	13.1	9.8
Montane chaparral	23.6	49	44	39	34	34	0.3	1.3	4.9	6.6	13.1	9.8
Montane hardwoods	36.4	75	67	60	52	52	0.3	1.3	4.9	6.6	13.1	13.1
Montane hardwoods conifer	25.8	54	48	43	37	37	0.3	1.3	4.9	6.6	13.1	9.8
Montane riparian	16.7	35	32	28	25	25	0.3	1.3	4.9	6.6	13.1	13.1
Perennial grass	6.4	53	32	21	11	11	0.3	1.3	4.9	6.6	13.1	6.6
Pinyon-juniper	13.9	71	50	36	21	21	0.3	1.3	4.9	6.6	13.1	9.8
Sagebrush	8.4	44	26	18	9	9	0.3	1.3	4.9	6.6	13.1	9.8
Sierran mixed conifer	39.9	82	74	65	57	57	0.3	1.3	4.9	6.6	13.1	13.1
Subalpine conifer	24.5	51	46	41	36	36	0.3	1.3	4.9	6.6	13.1	9.8
Urban	6.9	71	43	29	14	14	0.3	1.3	4.9	6.6	13.1	9.8
Valley foothill riparian	24.2	50	45	40	35	35	0.3	1.3	4.9	6.6	13.1	13.1
Wet meadow	14.3	73	44	29	15	15	0.3	1.3	4.9	6.6	13.1	9.8
White fir	47.3	97	87	77	68	68	0.3	1.3	4.9	6.6	13.1	13.1

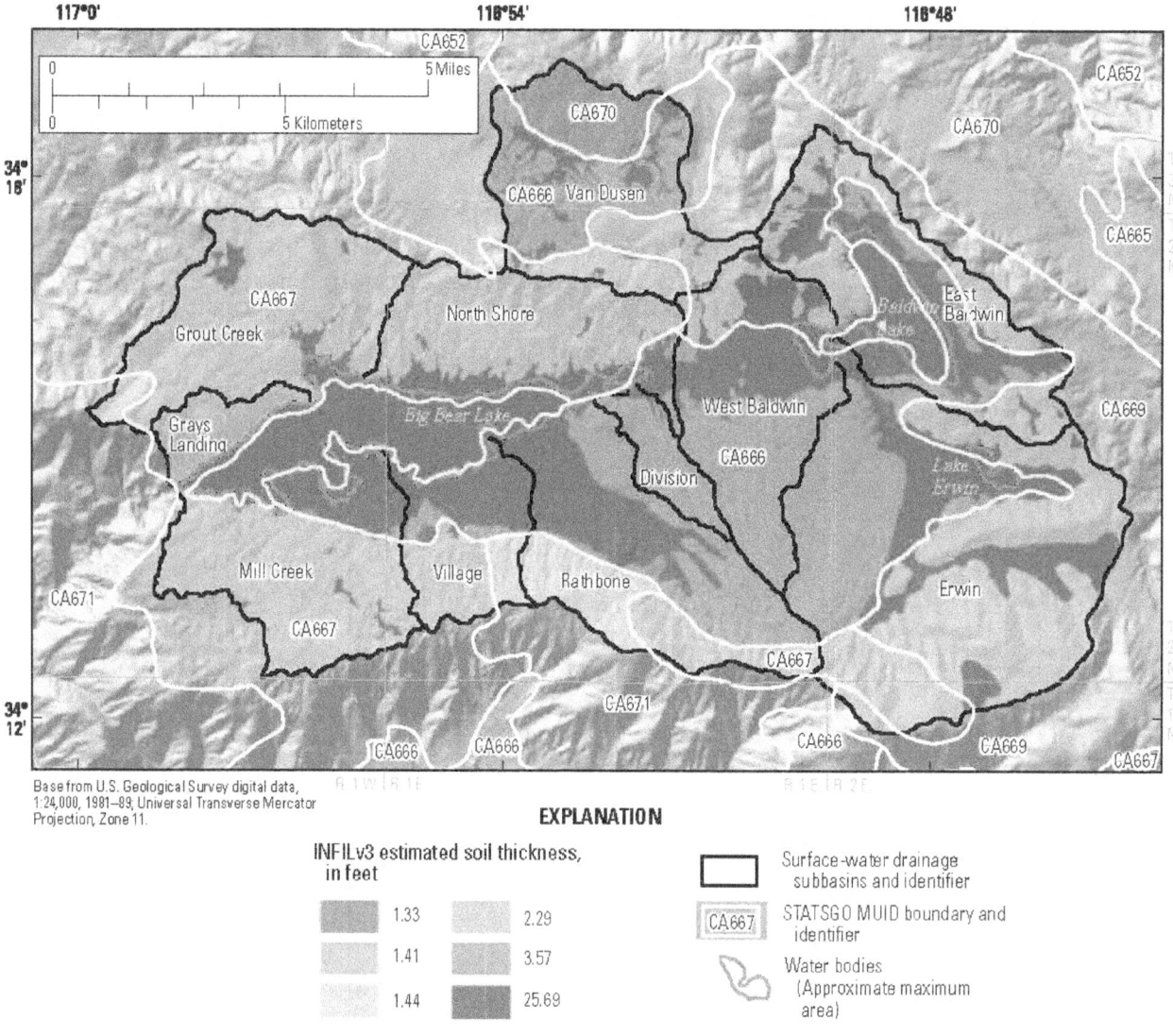

EXPLANATION

INFILv3 estimated soil thickness, in feet

[shade]	1.33	[shade]	2.29
[shade]	1.41	[shade]	3.57
[shade]	1.44	[shade]	25.69

☐ Surface-water drainage subbasins and identifier

CA667 STATSGO MUID boundary and identifier

Water bodies (Approximate maximum area)

Base from U.S. Geological Survey digital data, 1:24,000, 1981–89; Universal Transverse Mercator Projection, Zone 11.

Figure 23. STATSGO map-unit-identifier (MUID) and soil thickness map used in the INFILv3 model of Big Bear Valley, San Bernardino County, California.

Table 8. Soil parameters used in the INFILv3 model of Big Bear Valley, San Bernardino County, California.

STATSGO map unit identifier	Soil depth (feet)	Porosity	Field capacity	Wilting point	Brooks-Corey parameters		Saturated hydraulic conductivity (feet/day)
					Air-entry potential (bars)	Drainage coefficient	
CA666	3.57	0.397	0.186	0.041	−0.011	4.54	2.24
CA667	1.41	0.351	0.157	0.03	−0.01	4.21	2.47
CA669	2.29	0.436	0.163	0.045	−0.019	6.3	0.77
CA670	1.33	0.398	0.202	0.056	−0.015	5.47	1.28
CA671	1.44	0.426	0.121	0.014	−0.009	3.54	3.2

Model Coefficients

Model coefficients for simulating snowmelt and sublimation were identical to those used by Hevesi and others (2003). Precipitation was calculated for each grid cell and was assumed to be in the form of snow when the average daily air temperature was equal to or less than 32°F. Daily snowfall was added to the snowpack storage term in the daily water balance. When the average daily air temperature was less than or equal to freezing, the snow-cover term was reduced by a fraction defined by using an assumed sublimation model that calculated sublimation as a percentage of potential evapotranspiration and the available water in the snowpack. When the daily maximum air temperature was greater than freezing, an empirical temperature-index model was applied by using parameters calibrated for the Sierra Nevada (Maidment, 1993) to calculate the daily snowmelt, and the snowpack was reduced by this amount.

Model coefficients used to define average monthly atmospheric conditions needed for simulating potential evapotranspiration were the same as those used in Rewis and others (2006). Model coefficients used to represent stream channel characteristics included (1) the minimum number of upstream cells used to define the main stream channels and (2) the saturated hydraulic-conductivity multiplier for soils in the main stream channels. The minimum number of upstream cells was set to 100 (approximately 22.2 acres), and the saturated hydraulic-conductivity multiplier was set 10. This configuration assumed coarser soils in active channels with upstream areas of 22.2 acres or greater and a 10-fold increase in the saturated hydraulic conductivity of the channel bed relative to the surrounding inter-channel areas.

Boundary Conditions

Boundary conditions for the INFILv3 model were defined by the simulated daily surface-water discharge from all upstream model units that are direct tributaries to the model unit being simulated. To establish the boundary conditions, model units are simulated sequentially, starting with the upstream model units and following with downstream model units according to the routing order defined by the drainage network. The simulated surface-water discharges from an upstream model unit are input to the downstream model unit as daily inflows to the grid cell directly downstream of the outflow cell in the upstream unit.

The drainage network defining the Big Bear Lake surface-water drainage basin consists of 31 model units (*table 4*). All model units, except for model unit 16 (Division subbasin), are upstream tributaries to model unit 2 (the Big Bear Lake subbasin) and discharge directly into the Big Bear Lake subbasin (*table 4*). In the case of the Division subbasin, simulated daily outflow is discharged into the North Shore subbasin, which, in-turn, discharges into the Big Bear Lake subbasin. The Big Bear Lake subbasin collects all simulated outflows discharging into Big Bear Lake, including all streamflow from the tributary subbasins as well as runoff generated within the land areas of the Big Bear Lake subbasin.

The drainage network for the Baldwin Lake surface-water drainage basin consists of 9 model units connected in a series of linked subbasins and tributary model units (*table 4*). Simulated surface-water discharge from model unit 4 in the Erwin Lake subbasin defines the inflow boundary condition for downstream model unit 6, and simulated discharge from unit 6 is the inflow boundary condition for downstream model unit 27. The discharge from model unit 27 is the inflow boundary condition for model unit 7 and, also, is the surface-water outflow for the Erwin Lake subbasin into the West Baldwin subbasin. The West Baldwin subbasin also receives inflow from the upstream Van Dusen subbasin (model unit 5). The West Baldwin subbasin discharges into model unit 1, which is part of the East Baldwin subbasin. Within the East Baldwin subbasin, simulated surface-water discharges from model units 0, 1, and 26 define the inflow boundary conditions for model unit 3, which discharges into Baldwin Lake.

Initial Conditions

Initial conditions required by INFILv3 include the water contents of all root-zone layers, the perched zone (layer 7), and the snowpack. All simulations in this study were run by using an initial water content for root-zone layers 1 through 5 (soil layers) assumed to be 1.5 times the wilting point water content (*table 8*). An initial water content of zero was assumed for root-zone layer 6, the perched zone (layer 8), and the snowpack. Precipitation-runoff models, including INFILv3, generally require at least some initialization period in order to help minimize uncertainties associated with the assumed or estimated initial conditions. A one-year initialization period is adequate for snowpack storage for locations where snow cover does not persist through the summer months, such as the Big Bear area. Previous INFILv3 applications for a nearby study area indicated that a 2- to 3-yr initialization period for the root-zone water content was sufficient to generate results independent of the initial conditions assumed for most locations (Rewis and others, 2006).

The modified INFILv3 model used in this study required a longer initialization period of approximately 10 years to establish the ambient long-term average water contents of the SGWZ. Grid cells having the lowest-permeability bedrock assigned to the SGWZ required the longest initialization periods. The longer initialization period also was needed to establish the ambient longer-term seepage flows for the main stream channels. The length of the initialization period was determined using a trial-and-error method and is approximate because initialization was found to be dependent on several model parameters (particularly parameters defining the properties of the SGWZ). The significance of the length of the model initialization period to model application is discussed in the results sections that follow.

INFILv3 Model Calibration

Model calibration is the process of making adjustments, within justifiable ranges, to initial estimates of selected model parameters to obtain reasonable agreement between simulated and measured values. Precipitation-runoff models, such as INFILv3, usually are calibrated by comparing simulated streamflow to available records of measured streamflow, preferably using continuous records that span multi-year periods. Streamflow data are sparse for the Big Bear and Baldwin Lakes surface-water drainage basins, consisting of only a few measurements that correspond to when water-quality sampling was done and some annual peak-flow measurements (National Water Information System; waterdata.usgs.gov). Although streamflow data are sparse, multi-year records of lake levels for Big Bear Lake and Baldwin Lake are available.

In order to apply the lake-level records for model calibration, a daily water-balance model (referred to herein as the LAKE model) was developed to simulate lake levels and volumes for both Big Bear and Baldwin Lakes by using daily simulation results from the INFILv3 model. The daily inputs simulated by INFILv3, and used as input for the LAKE model, are precipitation, air-temperature, potential evapotranspiration, streamflow, and recharge. The LAKE model allows for the indirect calibration of the INFILv3 model by using a comparison of simulated and measured lake levels and volumes. The calibration of the INFILv3 model is indirect because, in addition to the INFILv3 model parameters, parameters in the LAKE model also are estimated and then adjusted during calibration.

Model calibration consisted of defining a single INFILv3 model for the entire study area by using a consistent set of INFILv3 model parameters and climate inputs for both Big Bear Lake and Baldwin Lake surface-water drainage basins. A best-fit INFILv3 model, referred to as the "base-case" model configuration, was calibrated to the Big Bear Lake volumes by using a trial-and-error process of varying INFILv3 model parameters, and then varying the LAKE model parameters for a given INFILv3 model configuration. Attempts were then made to calibrate the base-case model to Baldwin Lake volumes by trial-and-error adjustment of the LAKE model parameters for Baldwin Lake. If the calibration criteria could not be satisfied by using realistic parameters to define the LAKE model for Baldwin Lake, adjustments were made to the INFILv3 model configuration, and the process was repeated, starting with a re-calibration of the LAKE model for Big Bear Lake. Using this procedure, a final model calibration was used to define a single INFILv3 model configuration for the study area, allowing for differences in the parameters defining the separate LAKE models for Big Bear and Baldwin Lakes.

Model calibration of the combined INFILv3 and LAKE models was done by using graphical comparison of simulated and measured lake volumes and by evaluating three goodness-of-fit statistics: the Percent Average Estimation Error (PAEE), the Nash-Sutcliffe Model Efficiency (NSME), and the R-squared (r^2) from standard linear regression. The PAEE

is a measure of bias in the estimation error, and has a value of 0.0 percent for a purely unbiased model fit, such as the sample mean. The NSME is a standardized mean-square error statistic that is often used to compare results between different models (Nash and Sutcliffe, 1970). The NSME for the sample mean is 0.0; values less than 0.0 indicate a poor model fit relative to the sample mean, and values close to 1.0 indicate a good match between predicted and observed values (a value of 1.0 indicates a perfect fit). For this study, the following criteria were used to define a satisfactory model fit: absolute PAEE less than or equal to 10 percent, NSME greater than or equal to 0.5, and r^2 greater than or equal to 0.5.

Description of the LAKE Model

The LAKE model calculates the daily water balance for the maximum lake areas for Big Bear and Baldwin Lakes. The maximum lake area was estimated by using NHD data (*fig. 19*). Lake levels measured at Big Bear and Baldwin Lakes change with time; therefore, the wetted area, the dry-lakebed area, and the volume of the lakes also change with time. As part of the daily simulation, the LAKE model estimates the wetted-lake area and the dry-lakebed area on the basis of the simulated lake volume and a known or estimated lake level- area-volume relation, where the sum of the wetted-lake area and dry-lakebed area is equal to the maximum lake area. Initial conditions for the LAKE model are the lake volume and the soil-zone storage for the dry lakebed at the start of the simulation. The soil-zone storage accounts for the combined storage of water retention on the land surface and shallow subsurface storage in the soil of the lakebed.

LAKE Boundary Conditions

Simulation results from the INFILv3 model are used to define boundary conditions for the LAKE model on a daily basis: (1) precipitation, (2) air temperature, (3) potential evapotranspiration, (4) surface-water discharge (streamflow), and (5) groundwater discharge. Daily precipitation and streamflow are added directly as inflows to the LAKE model domain. Daily groundwater discharge is estimated using the INFILv3 simulated daily basin-wide recharge. The long-term average total groundwater-discharge rate is assumed to be either equal to or less than the long-term average recharge rate simulated by INFILv3. The total daily groundwater discharge is partitioned into a steady-state and a transient discharge component. The steady-state component is a constant daily groundwater-discharge rate applied to the length of the calibration period. The transient component uses a specified time-averaging period to calculate a time-averaged discharge rate on the basis of the daily recharge rate. Five different yearly time-averaging periods were defined: 1-, 2-, 3-, 5-, and 10-years. The fraction of total groundwater discharge partitioned into the steady-state and transient components was estimated and then adjusted during the LAKE model calibration. The time-averaging period providing the best calibration result also was identified using trial-and-error model fitting. The combined steady-state and transient discharge components were defined such that the total long-term average groundwater discharge rate was equal to or less than the long-term average INFILv3 recharge rate.

The LAKE model defined for Big Bear Lake assumed that all of the INFILv3 simulated recharge in the Big Bear Valley surface-water drainage basin discharges to the area of the lake. In the Baldwin Lake surface-water drainage basin, groundwater-discharge areas include the area of Baldwin Lake and wetted areas (water bodies and wetlands) upstream of the lake area. The groundwater-discharge area for the LAKE model defined for Baldwin Lake was estimated using the NHD mapped hydrographic features and aerial photographs. The NHD hydrographic features indicated three wetted areas upstream of Baldwin Lake: (1) Lake Erwin (103 acres), (2) an area of wetlands downstream of Lake Erwin (34 acres), and (3) Deadmans Lake (7 acres) upstream of Lake Erwin (*fig. 1*). In addition to the mapped hydrographic features, about 96 acres along the shoreline of Baldwin Lake were added to the total groundwater-discharge area to account for observed areas of seepage and spring discharge adjacent to the Baldwin Lake boundary. The total groundwater-discharge area (about 240 acres) was treated in the LAKE model as a single area. The discharge area was reduced by 20 percent to account for the potential evapotranspiration energy already applied by INFILv3 to simulate land-surface evapotranspiration.

The LAKE model for Baldwin Lake accounts for evapotranspiration losses from the upstream groundwater-discharge areas. The daily potential-evapotranspiration rate simulated by INFILv3 was used to simulate the evapotranspiration losses on the basis of the estimated effective groundwater-discharge area and the available daily groundwater discharge. The available daily groundwater discharge is determined by the INFILv3 simulated recharge in the Baldwin Lake surface-water drainage basin. The LAKE model simulates evapotranspiration occurring at its full potential (saturated conditions exist) in the groundwater-discharge areas. If the estimated daily groundwater discharge is greater than or equal to the maximum daily potential evapotranspiration, the groundwater discharge to the lake area is decreased by the amount of potential evapotranspiration. If the estimated daily discharge is less than the maximum daily potential evapotranspiration, groundwater discharge to the lake area is zero.

LAKE Water Balance For The Dry Lakebed

After estimating the groundwater discharge to the lake area, the LAKE model estimates the water balance for the dry-lakebed area of the lake. Daily inflows to the dry lakebed include the INFILv3 simulated surface-water discharge (streamflow) from surrounding drainages, direct precipitation on the dry lakebed, and groundwater discharge. The surface-water discharge includes the overland runoff and the seepage components for all model units (land areas) upstream of the lake area. Precipitation falling as snow on the dry lakebed was allowed to accumulate using a snowpack storage term. Daily sublimation and snowmelt losses from the snowpack were simulated using potential evapotranspiration and air temperature simulated by INFILv3. A fraction of precipitation falling as rain is routed as Hortonian runoff directly to the wetted-lake area.

The partitioning of surface-water and groundwater inflows between the dry lakebed and wetted-lake area is defined as a function of the dry-lakebed area. The fraction of surface-water inflow added to the dry lakebed is a function of the ratio of the dry-lakebed area to the maximum lake area. The quantity of groundwater inflow (the sum of the steady state and transient components) also is a function of the ratio of the dry-lakebed area to the maximum lake area. The function includes parameters defining the minimum fraction of surface-water and groundwater inflow to the wetted-lake area. The minimum fractions of surface-water and groundwater inflow were estimated and then adjusted as part of the calibration process. The inflows to the dry lakebed are retained in a combined surface-retention and shallow soil-storage term that allows evapotranspiration of available water from storage. Total runoff from the dry-lakebed area to the wetted-lake area is equal to the quantity of surface-water discharge from surrounding drainages, rain, snowmelt, and groundwater discharge exceeding the surface retention and soil-storage capacity of the dry lakebed, plus the direct Hortonian runoff component. The quantity of runoff was calculated for each daily time step and added to the lake volume.

LAKE Water Balance For The Wetted Area

After estimating the water balance for the dry-lakebed area of the lake, the LAKE model estimates the water balance for the wetted area of the lake. The model calculates an updated volume for the lake during each time step. Inflows to the wetted-lake area include precipitation falling directly on the wetted-lake area, surface-water runoff from the dry lakebed, surface-water discharge from the surrounding drainages, and groundwater discharge. INFILv3 simulates daily precipitation on the wetted area and the total potential surface-water discharge from the surrounding drainages. The LAKE model, as described in the "LAKE water balance for the dry lakebed" section, calculates runoff from the dry lakebed to the wetted area. Total potential groundwater discharge is less than or equal to the recharge simulated by INFILv3, as described in the "LAKE boundary conditions" section of this report. The actual surface-water and groundwater discharge added directly to the wetted area is the total discharge minus the inflow to the dry lakebed. All precipitation (both rain and snow) over the wetted area was added directly to the liquid water volume; freezing and thawing of the lake surface, and the accumulation, sublimation, and melting of snow on a frozen lake surface were not represented in the model.

Outflow from the wetted-lake area includes evaporation from the lake surface and surface-water discharge from the wetted-lake area. Evapotranspiration from the wetted area was equal to the INFILv3 simulated daily potential evapotranspiration, averaged over the wetted area. The surface-water discharge was calculated using a stage-discharge relation. Outflow through the lake bottom was assumed to be zero. After accounting for all inflows and outflows to the wetted area, the new lake volume was calculated and used as the initial lake volume for the next time step.

LAKE Model Parameters

The LAKE model parameters adjusted as part of the calibration process were (1) the steady-state and transient groundwater-inflow fractions, (2) the transient groundwater averaging period (1, 2, 3, 5, or 10 years), (3) the minimum fraction of surface-water inflow to the wetted area, (4) the minimum fraction of groundwater inflow to the wetted area, (5) the maximum water-storage capacity of the soil zone, (6) the Hortonian runoff fraction, (7) the initial water content of the soil zone, and (8) the initial water volume of the wetted area. If a satisfactory calibration could not be achieved by adjusting these eight parameters, the inflow boundary conditions simulated by INFILv3 were adjusted using multipliers for precipitation, surface-water discharge, and recharge (the multipliers were initially set to 1.0 for calibration). The multipliers were used to evaluate the sensitivity of the simulated lake volumes to changes in the relative magnitude of the INFILv3 simulated inflows. In addition, the multipliers for surface water and groundwater were used to evaluate potential outflows from the Baldwin Lake surface-water drainage basin. Decreasing surface-water or groundwater discharges to the lake area is appropriate under the assumption that the drainage basin might not be closed and, therefore, surface-water and groundwater losses occur across the basin boundaries. Daniel B. Stephens & Assoc. reported estimates of outflows from the southeastern part of the drainage basin of approximately 300–1,300 acre-ft/yr (Steve Cullen, D.B. Stephens and Assoc., written communication, February 25, 2009), providing evidence that the Baldwin Lake surface-water drainage basin might not be a closed basin.

Calibration Targets

Measured lake volumes for Big Bear and Baldwin Lakes were used as calibration targets for the INFILv3 model of Big Bear Valley. Calibration using lake volumes was preferred to calibration using lake levels because lake volumes are more directly related to the water-balance simulation (similar to the use of stream discharge rather than stream stage for model calibration).

Measured Lake Levels And Volumes for Big Bear Lake

Lake-level data for Big Bear Lake were obtained from the MWD (*www.bbmwd.com/Lake_Intro.html*, accessed January 2007) for approximately weekly intervals from January 1, 1985, through May 26, 2006. Lake-level data also were available from October 18, 2004, through May 26, 2006. A lake level of 72 ft corresponds to lake volume of 72,358 acre-ft, which is the maximum volume, and 55 ft corresponds to a volume of 29,586 acre-ft, which is the minimum volume. The lake-level data and two volume measurements were used to develop a linear relation between level and volume. Lake surface area then was estimated as a linear function of lake volume between the maximum and minimum lake volume. At a lake level of zero, corresponding to an altitude of 6,669.75 feet above mean sea level, minimum volume and area were extrapolated to zero. The relations between level, volume, and surface area were used to develop a rating table that was input to the LAKE model. The estimated volumes for Big Bear Lake were calculated using the rating table and approximately weekly measured lake levels for January 1, 1985, through October 17, 2004. Measured lake levels and volumes were available from October 17, 2004, through December 26, 2005, and were also used as input to the LAKE model. The estimated record of lake volumes for January 1, 1985, through December 26, 2005, was then used for calibration. There were 1,145 lake-volume estimates; the mean lake volume for this period was 56,049 acre-ft, the maximum volume was 73,000 acre-ft, and the minimum volume was 29,548 acre-ft (*table 9*).

The rating table used in the LAKE model for Big Bear Lake also included an estimated stage-discharge relation for simulating surface-water discharge at Bear Valley Dam. A constant discharge of 0.3 cubic feet per second (ft³/s), about 0.6 acre-ft/day, was used for lake altitudes from 6,727 ft (equal to a lake volume of 34,400 acre-feet, or 47 percent of the full storage capacity) to 6,735 ft (a lake volume of 52,600 acre-feet, or 72 percent of the full storage capacity). The constant discharge of 0.3 ft³/s at these low lake levels is the minimum flow needed to maintain fish habitat in the natural stream channel downstream of the dam (Big Bear Municipal Water District, personal commun., January 2004). Below 6,727 ft, discharge was assumed to decrease steadily to zero at an altitude of 6,712 ft (a lake volume of 13,000 acre-ft). Above 6,735 ft, discharge was assumed to steadily increase to 50 ft³/s at a lake altitude of 6,742 ft (corresponding to a lake depth of 72.33 ft and a full storage capacity at 73,000 acre-ft). Above 6,742 ft, discharge was assumed to increase more rapidly to 2,000 ft³/s at a lake altitude of 6,745 ft, and then to 5,000 ft³/s at an altitude of 6,748 ft, accounting for much higher discharges through the spillway.

Table 9. Observed lake levels and calculated lake areas and volumes used for model calibration of the LAKE models for Big Bear and Baldwin Lakes, Big Bear Valley, San Bernardino County, California.

Parameter	Units	Big Bear Lake	Baldwin Lake
Period of record used for calibration	Month/day/year	01/01/1985 to 12/26/2005	10/01/1949 to 08/20/1999
Approximate observation frequency	Not applicable	Weekly	Monthly
Type of observation	Not applicable	Instantaneous	Average monthly
Number of observations	Not applicable	1,145	599
Observed mean lake attitude	Feet	6,736.80	6,697.50
Observed maximum lake atitude	Feet	6,743.20	6,706.80
Observed minimum lake atitude	Feet	6,725.70	6,695.30
Observed mean lake level	Feet	66	2.5
Observed maximum lake level	Feet	72.3	11.8
Observed minimum lake level	Feet	54.9	0.3
Calculated mean lake area	Acres	2,192	274
Calculated maximum lake area	Acres	2,854	723
Calculated minimum lake area	Acres	1,155	153
Calculated mean lake volume	Acre-feet	56,049	731
Calculated maximum lake volume	Acre-feet	73,000	5,283
Calculated minimum lake volume	Acre-feet	29,548	43

Measured Lake Levels And Volumes for Baldwin Lake

Lake-level data were recorded for Baldwin Lake on a periodic basis from June 1934 through August 1999 on a hard-copy chart archived by the Big Bear City Community Services District (Big Bear City Community Services District, written commun., 2005). These data were recorded as absolute lake altitudes (altitude above mean sea level) on the chart, and visual inspection of the chart record indicated that lake altitudes were measured approximately three times a year (April, August, and December). Lake altitudes less than approximately 6,695.3 to 6,695.5 ft were represented by a dotted line on the chart record. The dotted line was interpreted as indicating that the lake altitude was less than the altitude of the gage during these observation dates. The chart record was electronically digitized to develop an approximate record of average monthly lake altitudes.

A rating table for the lake level–area–volume relation for Baldwin Lake was developed by using the 30-m DEM also used to develop the INFILv3 model. The DEM indicated a closed depression characteristic of a dry playa lake and was assumed to approximately represent the lake bathymetry. The lowest consistent lakebed altitude for the DEM was 6,692.9 ft NAVD, although some isolated areas within the Baldwin Lake playa indicated slightly lower altitudes of 6691.6 ft. The chart record indicated a minimum playa lakebed altitude of 6695.0 ft. In order to make the lake-altitude data consistent with the lake level-area-volume relation developed from the DEM,

the lake-altitude data were shifted down by 2.1 ft (the difference between the minimum altitude on the chart record and the minimum lakebed altitude of the rating table). A maximum lakebed area of 1,079 acres was defined using the NHD boundary for Baldwin Lake (the maximum area includes the area associated with the sewage-treatment ponds). The stage-discharge relation developed for Baldwin Lake assumed zero surface-water discharge for altitudes less than 6,745 ft, and allowed for spill-over of Baldwin Lake into Big Bear Lake for altitudes greater than 6,745 ft.

Monthly lake volumes for Baldwin Lake were calculated from October 1, 1949, through August 20, 1999, by using the developed lake-area-volume relation and the digitized record of measured lake altitudes. On the basis of the lake level-area-volume relation developed for this study, combined with the −2.1 ft shift applied to the lake-level record, there could be a maximum of 43 acre-ft in Baldwin Lake when the lake level is reported at its lowest altitude (6,695.3 ft NAVD). Baldwin Lake-level records prior to October 1, 1949, were not included in model calibration because of uncertainty in model results due to sparse precipitation and air-temperature records prior to water year 1949. The developed October 1, 1949–August 20, 1999, record of average monthly lake levels, areas, and volumes were developed for the period from October 1, 1949 to August 20, 1999; the mean lake volume was 731 acre-ft, the maximum volume of 5,283 acre-ft, and the minimum volume was 43 acre-ft (*table 9*).

Base-Case Model Calibration

An initial INFILv3 model, herein referred to as the base-case model, was developed by using the preliminary climate model as input. As described in the "Climate input" section of this report, the preliminary climate model was developed by using average monthly PRISM precipitation maps for the 30-year period of 1971–2000 in combination with daily precipitation and air-temperature records from 35 NCDC stations. Calibration of the INFILv3 model was achieved by adjusting (1) soil thickness, (2) the hydraulic conductivity of root-zone layer 6 (including bedrock and unconsolidated deposits), (3) the estimated thicknesses of root-zone layer 6, (4) the estimated storage capacity of the perched zone, (5), the seepage factor (a multiplier used to increase or decrease the simulated seepage rate), and (6) the coefficients defining estimated root density for each root zone. The initial INFILv3 parameter values used in the base-case model were similar to values estimated in previous applications of INFILv3 in southern California (Rewis and others, 2006; Nishikawa and others, 2004). The initial parameter values were adjusted until simulated volumes in both models (INFILv3 and LAKE) approximated estimated volumes in Big Bear Lake. Calibration was done by manual trial-and-error adjustment of selected INFILv3 and LAKE model parameters to improve the match between simulated and measured lake volumes.

Base-Case Model Calibration For Big Bear Lake

The base-case model was considered calibrated when simulated Big Bear Lake volumes approximated measured volumes. The INFILv3 base-case model simulated average annual water balance for water years 1950–2005 for the Big Bear Lake surface-water drainage basin were compiled by subbasin (table 10A). The simulated average annual precipitation was about 44,170 acre-ft/yr, and about 73 percent of the simulated precipitation in the basin was consumed by evapotranspiration. The simulated average annual recharge was about 4,030 acre-ft/yr (about 9.1 percent of precipitation), and the average annual surface-water outflow was about 5,420 acre-ft/yr (about 12.3 percent of precipitation; table 10). The ratio of recharge to net surface water outflow was about 0.7 for the Big Bear Lake surface-water drainage basin, indicating a runoff-dominated basin.

The INFILv3 simulated surface-water outflow and recharge were used as inflows to the LAKE model for Big Bear Lake, where surface-water outflow is input as surface-water runoff in the LAKE model and recharge is input as groundwater discharge (table 11). A visual comparison of the simulated and measured lake volumes for the calibration period 1985–2005 indicates a good match (fig. 24A). The goodness of fit statistics also indicate a good model fit, with a PAEE of 0.15 percent, a NSME of 0.95, and a regression coefficient (r^2) of 0.95 (table 11). The best fit to measured lake volumes was achieved using a 2-year running average of the INFILv3 simulated daily upstream recharge (representing the transient groundwater-inflow component) and 10 percent of the long-term average INFILv3 simulated basin-wide recharge (about 403 acre-ft/yr; representing the steady-state groundwater-inflow component). Additional parameters defining the base-case model include a value of 1.0 for the minimum fraction of surface-water inflow to the wetted-lake area, a value of 0.25 for the minimum fraction of groundwater inflow to the wetted-lake area, a maximum soil-zone storage capacity of 20 in., an initial soil-zone water content of 5 in., and a Hortonian runoff fraction of 0.6 (table 11).

The average simulated Big Bear Lake volume for calendar years 1985–2005 was 56,135 acre-ft (table 11), which compared well with the average estimated lake volume of 56,049 acre-ft (table 9). The period of maximum measured lake volumes during the mid-to-late 1990s also was reproduced well by the model, as were seasonal fluctuations in lake volumes (fig. 24A). The maximum simulated lake volume of 76,533 acre-ft occurred in water-year 1995 (this volume indicates flooding conditions), which corresponded to several estimated maximum volumes that occurred during the calibration period. The average simulated lake level was 66.2 ft, compared with a measured average level of 66.0 ft (tables 9 and 11). Overall, the difference between simulated and measured lake levels and volumes was less than about 3 percent, except for water-years 2002–04, during which time the model overestimated lake volumes (fig. 24A). This difference was probably caused by assuming that the total quantity of INFILv3 simulated recharge is available for groundwater discharge to the lake. Groundwater pumping, however, would reduce the quantity of groundwater discharge to the lake. Groundwater pumping, which averaged over 2,500 acre-ft/yr during this period (William S. La Haye, Water Resource Manager, Big Bear Lake DWP, written commun., 2010), could account for the difference between the simulated and observed lake volumes.

Annual simulation results of the base-case model for entire simulation period (water-years 1928–2005) indicate a high degree of year-to-year variability in the simulated inflows, outflows, and changes in lake volume for Big Bear Lake (fig. 24B). The greatest variability occurred in the simulated surface-water inflows and outflows. High annual surface-water inflows, greater than approximately 20,000 acre-ft/yr, were simulated for water years 1969, 1978, 1980, and 2005 (fig. 24B). A maximum annual outflow of about 32,000 acre-ft and a maximum annual surface-water inflow of about 28,000 acre-ft were simulated for water-year 1969. For the drier years, surface-water inflows were greater than outflow at Bear Valley Dam because of lake evaporation. The simulated groundwater inflow usually exceeded surface-water inflow for the drier years.

Table 10. INFILv3 simulated average annual water-balance for the base-case model of the Big Bear and Baldwin Lakes surface-water drainage basins, Big Bear Valley, San Bernardino County, California, water-years 1950–2005.

[Abbreviations: $\geq$, greater than or equal to; <, less than]

| Subbasin | Simulated basin and subbasin water balance (acre-feet per year) | | | | | | | | | | | Shallow groundwater zone Average grid-cell balance (inch per year) | | | | Recharge indicators | | |
| | Inflow | | | | Change in storage (water content) | | Outflow | | | | Basin-wide average water content (acre-feet) | Change in water content | Inflow | Outflows | | | | |
	Total precip-itation	Rainfall	Snow fall	Surface-water inflow	Root zone	Shallow ground-water zone	Subli-mation	Evapo-transpira-tion	Surface-water outflow	Recharge			Net infiltration	Seepage	Recharge	Ratio of recharge to net outflow [1]	Ratio of seepage to recharge	Recharge as pecent of precipi-tation
						Big Bear Lake surface-water drainage basin												
Big Bear Lake	866	495	371	5,338	3	0.2	36	681	5,422	60	64	0.01	2.29	0.56	1.72	0.7	0.3	6.9
Grout Creek	10,012	4,452	5,559	0	6	8.4	571	6,832	2,097	498	1,665	0.02	2.42	1	1.39	0.2	0.7	5
Village	4,088	2,079	2,009	0	13.7	5	173	3,430	166	301	915	0.03	3.16	1.37	1.76	1.8	0.8	7.4
Division	1,456	810	646	0	2.1	7.5	59	1,283	4	101	228	0.11	3.71	2.07	1.53	26.3	1.3	6.9
Mill Creek	9,869	4,531	5,338	0	13.8	5.9	475	6,472	2,388	515	1,575	0.02	2.07	0.53	1.53	0.2	0.3	5.2
Rathbone	9,538	4,309	5,229	0	28.2	18.1	451	7,682	64	1,295	1,498	0.05	7.57	4.28	3.25	20.4	1.3	13.6
North Shore	6,528	3,338	3,190	4	8	6.3	366	4,817	73	1,262	543	0.02	8.64	4.17	4.45	18.3	0.9	19.3
Gray's Landing	1,814	778	1,036	0	0	0.6	118	1,142	551	2	116	0.01	0.14	0.1	0.03	0	3.5	0.1
Average/total	44,171	20,793	23,378	0	74.9	52.1	2,250	32,339	5,422	4,034	6,604	0.03	4.63	2.24	2.36	0.7	0.9	9.1
						Baldwin Lake surface-water drainage basin												
Van Dusen	8,727	3,722	5,004	0	19.9	6.7	535	6,762	213	1,190	1,311	0.02	5.48	2.18	3.29	5.6	0.7	13.6
West Baldwin	7,271	3,971	3,300	426	23.8	25.9	334	6,298	405	610	728	0.08	3.72	1.74	1.9	-28.8	0.9	8.4
Erwin	21,072	8,565	12,508	0	42.6	34.8	1,144	16,058	213	3,575	4,188	0.04	7.91	3.41	4.45	16.8	0.8	17
East Baldwin	6,953	3,839	3,114	405	21.3	2.5	306	5,909	500	619	1,603	0.01	2.77	0.86	1.9	6.5	0.5	8.9
Average/total	44,023	20,096	23,926	0	107.6	69.8	2,319	35,027	500	5,994	7,830	0.04	5.76	2.41	3.31	12	0.7	13.6
Study area total	88,194	40,890	47,305	0	182	122	4,569	67,366	5,923	10,028	14,433	0.03	5.21	2.33	2.85	1.7	0.8	11.4

[1] Absolute values $\geq$1.0 identify recharge dominated subbasins; absolute values <1 identify runoff or outflow dominated subbasins. Negative values indicate subbasins where surface-water inflow exceeds surface-water outflow.

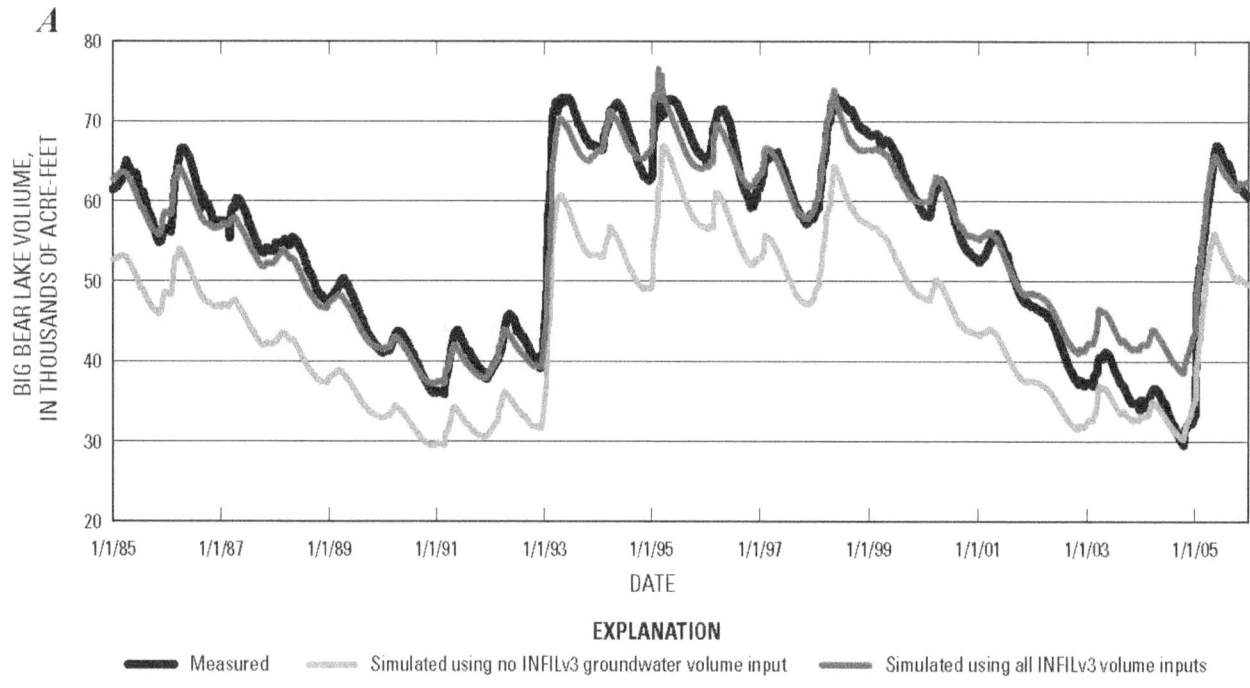

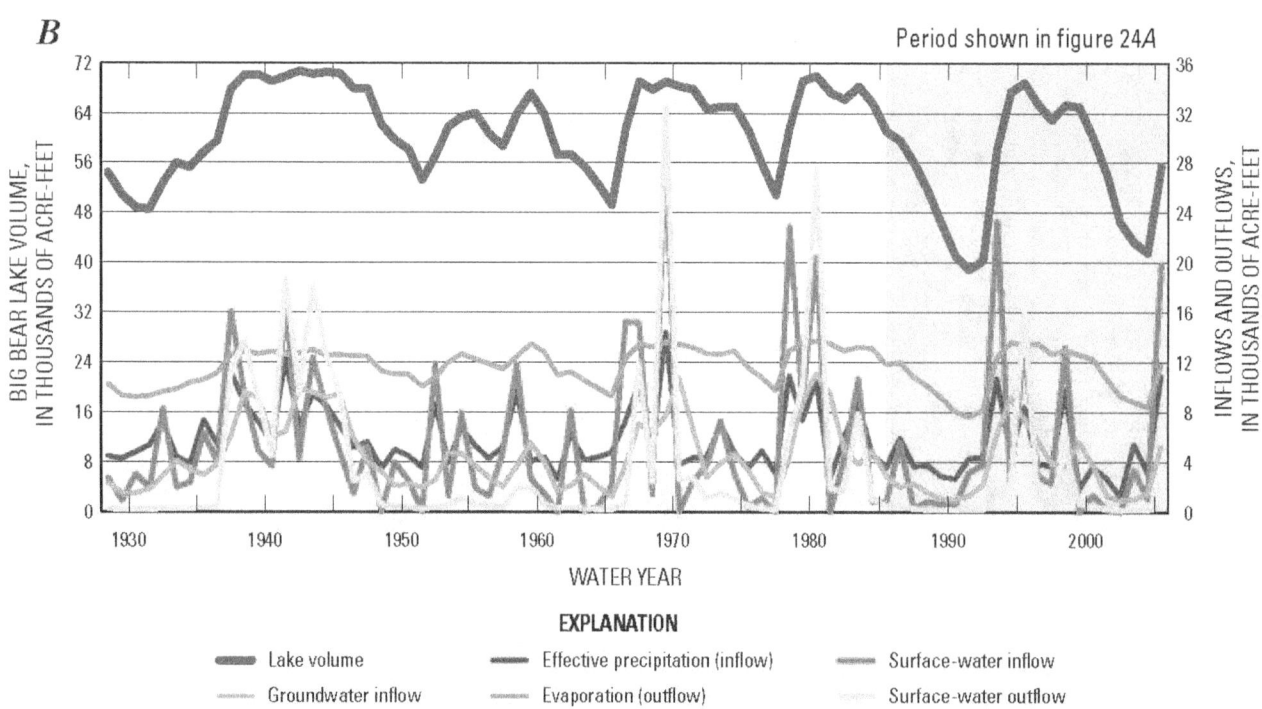

Figure 24. (*A*) Comparison of simulated and observed lake volumes for calendar years 1985–2005 and (*B*) simulated results for water years 1928–2005 for selected water-balance components of LAKE model of Big Bear Lake using the base-case INFILv3 model, Big Bear Valley, San Bernardino County, California.

Table 11. LAKE base-case model results for Big Bear Lake for calendar years 1985–2005, calibration statistics, and model parameters, Big Bear Valley, San Bernardino County, California.

Parameter	Units	INFILv3 base-case model, Big Bear Lake		INFILv3 base-case model, Baldwin Lake	
		Base-case model results and calibration statistics			
INFILv3 base-case model inflow volumes	None	No reduction in INFILv3 inflows	No groundwater inflow	No reduction in INFILv3 inflows	No groundwater inflow
Simulated mean lake level	Feet	66.1	62.1	46.3	2
Simulated maximum lake level	Feet	74.1	70	49.2	15.7
Simulated minimum lake level	Feet	58.4	54.8	44.5	0
Simulated mean lake volume	Acre-feet	56,135	46,070	57,100	733
Simulated maximum lake volume	Acre-feet	76,533	66,826	63,502	8,555
Simulated minimum lake volume	Acre-feet	37,093	29,407	53,534	0
Percent average estimation error (PAEE)	Percent	0.15	−17.8	7,708.99	0.27
Nash-Sutcliffe Model Efficiency (NSME)	None	0.95	0.23	−2,948.56	0.31
R-squared	None	0.95	0.94	0.23	0.69
		LAKE model parameters (base-case model)			
Fraction of INFILv3 precipitation inflow	Decimal	1.00	1.00	1.00	1.00
Fraction of INFILv3 surface-water inflow	Decimal	1.00	1.00	1.00	1.00
Fraction of INFILv3 groundwater inflow	Decimal	1.00	0.00	1.00	0.00
Steady state groundwater inflow fraction	Decimal	0.10	0.10	1.00	0.00
Transient groundwater inflow fraction	Decimal	0.90	0.90	0.00	0.00
Transient recharge averaging period	Years	2.00	2.00	0.00	0.00
Upstream groundwater discharge area	Acres	0.00	0.00	192.23	192.23
Minimum fraction surface-water inflow to wetted area	Decimal	1.00	1.00	0.00	0.00
Minimum fraction groundwater inflow to wetted area	Decimal	0.25	0.25	0.00	0.00
Maximum soil zone storage capacity	Inches	20.00	20.00	20.00	43.00
Initial soil zone storage capacity	Inches	5.00	5.00	5.00	5.00
Hortonian runoff fraction	Decimal	0.60	0.60	0.20	0.00
Initial lake volume	Acre-feet	56,049	56,049	731	731

The effects of groundwater inflow on the simulated lake volume of Big Bear Lake were also evaluated. Groundwater inflow was set to zero for all time steps in the calibrated LAKE model, which resulted in a poor match between estimated and simulated lake volumes (*fig. 24A*) and demonstrated the importance of groundwater inflow to the water balance of the lake. The simulated average lake volume for the base-case model for water-years 1950–2005 was 56,135 acre-ft/yr when groundwater inflow was simulated in the LAKE model; the simulated average lake volume only was 46,070 acre-ft/yr when groundwater inflow was not simulated (*table 11*). *Table 12A* shows a comparison of the LAKE model simulated 1950-2005 average inflows, outflows, and changes in storage for Big Bear Lake with and without groundwater inflow to the lake area.

Base-Case Model Calibration for Baldwin Lake

After satisfactorily calibrating the INFILv3 base-case model to Big Bear Lake estimated lake volumes, the INFIL3 base-case model inflows from the Baldwin Lake surface-water drainage basin were used to develop a LAKE model

for Baldwin Lake to test the calibrated model. The simulated average annual precipitation for water-years 1950–2005 was about 44,020 acre-ft/yr, with about 80 percent of the simulated precipitation in the basin consumed by evapotranspiration (table 10). The simulated average annual recharge was about 5,990 acre-ft/yr (about 13.6 percent of precipitation), and the average annual surface-water outflow was about 500 acre-ft/yr (about 1.1 percent of precipitation) (*table 10*). The ratio of recharge to net surface-water outflow was about 12 for the Baldwin Lake surface-water drainage basin, indicating a recharge-dominated basin.

Unlike the close model fit obtained with the Big Bear Lake calibration, simulated inflows from the base-case model for water-years 1950–99 did not provide a reasonable match to the Baldwin Lake volumes (*fig. 25A*). Results from trial-and-error adjustment of LAKE model parameters indicated that a good model fit using 100 percent of the inflows simulated by the INFILv3 base-case model was not possible. Simulated inflows to Baldwin Lake from the base-case model resulted in the formation of a large permanent lake with an average volume of about 57,050 acre-ft and an average outflow of about 1,960 acre-ft/yr (*table 12*).

Table 12. INFILv3 and LAKE simulation results for water years 1950–2005 by using the base-case model and alternative INFILv3 model configurations for Big Bear Lake surface-water drainage basin and Baldwin Lake surface-water drainage basin, Big Bear Valley, San Bernardino County, California.

[Abbreviations: ac-ft/yr, acre-feet per year; ft, feet; ft/day, feet per day; °F, degrees Fahrenheit; NSME, Nash-Sutcliffe Model Efficiency; PAEE, Percent average estimation error; RC, revised climate]

Model	INFILv3 seepage factor	INFILv3 precipitation factor	INFILv3 air temperature shift (°F)	INFILv3 spatially distributed climate input	INFILv3 coupled surface-water flow routing	Average soil thickness (ft)	Vertical saturated hydraulic conductivity (K) Layer 6 (ft/day)	Vertical saturated hydraulic conductivity (K) Layer 7 (ft/day)	Precipitation (ac-ft/yr)	Air temperature (°F)	Snow (ac-ft/yr)	Evapotranspiration (ac-ft/yr)	Surface-water inflow to lake (ac-ft/yr)	Recharge (ac-ft/yr)	Total inflow to lake from INFILv3 (ac-ft/yr)	Average lake volume (ac-ft/yr)	Average lake discharge (ac-ft/yr)	PAEE	NSME
1	1	1	1	Preliminary	Yes	2.6	0.83	0.17	44,171	44.8	21,455	32,339	5,422	4,034	9,456	59,414	2,778	0.2	0.95
2	10	1	1	Preliminary	Yes	2.6	0.83	0.17	44,171	44.8	21,455	33,083	5,675	3,116	8,791	57,729	2,405	-2.4	0.94
3	0.1	1	1	Preliminary	Yes	2.6	0.83	0.17	44,171	44.8	21,455	31,528	5,369	4,880	10,248	60,913	3,302	3	0.92
4	0	1	1	Preliminary	Yes	2.6	0.17	0.17	44,171	44.8	21,455	30,473	6,314	5,062	11,375	61,747	4,279	4.5	0.89
5	0	1	1	Preliminary	Yes	2.6	0.83	0.83	44,171	44.8	21,455	29,363	4,928	7,563	12,491	63,796	5,042	8.6	0.75
6	0	1	1	Preliminary	Yes	2.6	0.03	0.03	44,171	44.8	21,455	30,790	7,358	3,697	11,055	60,490	4,196	2.1	0.93
7	0	1	1	Preliminary	Yes	2.6	4.17	4.17	44,171	44.8	21,455	28,417	3,649	9,790	13,438	65,271	5,743	11.6	0.59
8	10	1	1	Preliminary	Yes	2.6	0.83	0.83	44,171	44.8	21,455	30,780	5,200	5,866	11,066	60,750	3,175	2.8	0.92
9	100	1	1	Preliminary	Yes	2.6	0.83	0.83	44,171	44.8	21,455	30,911	5,567	5,371	10,938	61,497	3,883	4	0.9
10	1	1	1	Preliminary	Yes	2.6	4.17	0.03	44,171	44.8	21,455	35,926	4,530	1,283	5,813	49,626	900	-16.6	0.32
11	1	1	1	Preliminary	Yes	2.6	1.06	0.21	44,171	44.8	21,455	32,382	4,657	4,815	9,471	59,627	2,723	0.5	0.93
12	10	1	1	Preliminary	Yes	3.1	1.06	0.11	44,171	44.8	21,455	34,767	5,112	1,892	7,003	53,173	1,438	-9.7	0.73
13	5	1	1	Preliminary	Yes	3.1	1.06	0.21	44,171	44.8	21,455	33,742	4,807	3,238	8,044	56,405	1,874	-4.7	0.88
14	1	1	1	Preliminary	No	2.6	0.83	0.17	44,171	44.8	21,455	30,917	9,107	1,781	10,888	60,106	4,242	1.4	0.93
15	1	0.5	1	Preliminary	Yes	2.6	0.83	0.17	22,086	44.8	10,727	19,161	919	672	1,592	15,376	14	-74.4	-11.62
16	1	0.6	1	Preliminary	Yes	2.6	0.83	0.17	26,503	44.8	12,873	22,324	1,552	1,100	2,653	22,796	58	-62.8	-7.99
17	1	0.7	1	Preliminary	Yes	2.6	0.83	0.17	30,920	44.8	15,018	25,198	2,330	1,657	3,987	32,549	145	-47.8	-4.23
18	1	0.8	1	Preliminary	Yes	2.6	0.83	0.17	35,337	44.8	17,164	27,784	3,248	2,348	5,596	44,345	312	-28.1	-0.87
19	1	0.9	1	Preliminary	Yes	2.6	0.83	0.17	39,754	44.8	19,309	30,151	4,285	3,147	7,432	53,229	1,170	-9.9	0.69
20	1	1.1	1	Preliminary	Yes	2.6	0.83	0.17	48,589	44.8	23,600	34,367	6,640	5,008	11,648	62,762	5,026	6.7	0.82
21	1	1.2	1	Preliminary	Yes	2.6	0.83	0.17	53,006	44.8	25,746	36,248	7,928	6,066	13,994	65,075	7,672	11.8	0.52
22	1	1.3	1	Preliminary	Yes	2.6	0.83	0.17	57,423	44.8	27,891	37,982	9,284	7,205	16,489	66,746	10,551	15.7	0.16
23	1	1.4	1	Preliminary	Yes	2.6	0.83	0.17	61,840	44.8	30,037	39,578	10,705	8,421	19,126	68,058	13,602	19	-0.26
24	1	1.5	1	Preliminary	Yes	2.6	0.83	0.17	66,257	44.8	32,182	41,044	12,186	9,708	21,894	69,009	16,837	21.4	-0.59
25	1	1	-7.2	Preliminary	Yes	2.6	0.83	0.17	44,171	37.6	33,461	31,552	5,238	4,516	9,754	63,145	3,881	7.1	0.78
26	1	1	-3.6	Preliminary	Yes	2.6	0.83	0.17	44,171	41.2	28,813	31,931	5,215	4,228	9,443	61,261	3,136	3.3	0.89
27	1	1	-1.8	Preliminary	Yes	2.6	0.83	0.17	44,171	43	25,411	32,127	5,288	4,126	9,414	60,354	2,918	1.6	0.93
28	1	1	1.8	Preliminary	Yes	2.6	0.83	0.17	44,171	46.6	17,198	32,598	5,553	3,967	9,520	58,577	2,651	-0.9	0.94
29	1	1	3.6	Preliminary	Yes	2.6	0.83	0.17	44,171	48.4	13,148	32,893	5,682	3,895	9,577	57,758	2,517	-2	0.93
30	1	1	7.2	Preliminary	Yes	2.6	0.83	0.17	44,171	52	6,766	33,531	5,896	3,712	9,608	55,850	2,231	-5	0.86
RC	1	1	1	Revised	Yes	2.6	0.83	0.17	38,574	44.8	20,058	29,702	3,989	2,798	6,788	50,368	915	-13.7	0.49

Table 12. INFILv3 and LAKE simulation results for water years 1950–2005 by using the base-case model and alternative INFILv3 model configurations for Big Bear Lake surface-water drainage basin and Baldwin Lake surface-water drainage basin, Big Bear Valley, San Bernardino County, California.—Continued

[Abbreviations: ac-ft/yr, acre-feet per year; ft, feet; ft/day, feet per day; °F, degrees Fahrenheit; NSME, Nash-Sutcliffe Model Efficiency; PAEE, Percent average estimation error; RC, revised climate]

| | INFILv3 parameters | | | | | | | | Baldwin Lake surface-water drainage basin — INFILv3 simulation results | | | | | | |
Model	INFILv3 seepage factor	INFILv3 precipitation factor	INFILv3 air temperature shift (°F)	INFILv3 spatially distributed climate input	INFILv3 coupled surface-water flow routing	Average soil thickness (ft)	Vertical saturated hydraulic conductivity (K) Layer 6 (ft/day)	Vertical saturated hydraulic conductivity (K) Layer 7 (ft/day)	Precipitation (ac-ft/yr)	Air temperature (°F)	Snow (ac-ft/yr)	Evapotranspiration (ac-ft/yr)	Surface-water inflow to lake (ac-ft/yr)	Total recharge (ac-ft/yr)	Total inflow to lake from INFILv3 (ac-ft/yr)
1	1	1	1	Preliminary	Yes	3	1.8	0.36	44,023	44.5	23,926	35,027	500	5,994	6,495
2	10	1	1	Preliminary	Yes	3	1.8	0.36	44,023	44.5	23,926	35,888	680	4,993	5,673
3	0.1	1	1	Preliminary	Yes	3	1.8	0.36	44,023	44.5	23,926	34,043	481	6,950	7,431
4	0	1	1	Preliminary	Yes	3	0.36	0.36	44,023	44.5	23,926	33,610	781	7,199	7,980
5	0	1	1	Preliminary	Yes	3	1.8	1.8	44,023	44.5	23,926	32,725	443	8,439	8,882
6	0	1	1	Preliminary	Yes	3	0.07	0.07	44,023	44.5	23,926	34,024	1,936	5,621	7,557
7	0	1	1	Preliminary	Yes	3	9.01	9.01	44,023	44.5	23,926	32,128	315	9,173	9,487
8	10	1	1	Preliminary	Yes	3	1.8	1.8	44,023	44.5	23,926	33,712	520	7,373	7,893
9	100	1	1	Preliminary	Yes	3	1.8	1.8	44,023	44.5	23,926	34,081	620	6,897	7,517
10	1	1	1	Preliminary	Yes	3	9.01	0.07	44,023	44.5	23,926	37,971	1,091	2,449	3,540
11	1	1	1	Preliminary	Yes	3	2.34	0.47	44,023	44.5	23,926	34,310	428	6,855	7,283
12	10	1	1	Preliminary	Yes	4.1	2.34	0.23	44,023	44.5	23,926	37,498	801	3,152	3,953
13	5	1	1	Preliminary	Yes	4.1	2.34	0.47	44,023	44.5	23,926	36,434	482	4,552	5,033
14	1	1	1	Preliminary	No	3	1.8	0.36	44,023	44.5	23,926	33,318	5,785	2,448	8,233
15	1	0.5	1	Preliminary	Yes	3	1.8	0.36	22,011	44.5	11,605	19,872	4	696	700
16	1	0.6	1	Preliminary	Yes	3	1.8	0.36	26,414	44.5	13,926	23,466	28	1,277	1,305
17	1	0.7	1	Preliminary	Yes	3	1.8	0.36	30,816	44.5	16,248	26,758	87	2,103	2,190
18	1	0.8	1	Preliminary	Yes	3	1.8	0.36	35,218	44.5	18,569	29,769	182	3,181	3,362
19	1	0.9	1	Preliminary	Yes	3	1.8	0.36	39,621	44.5	20,890	32,524	312	4,486	4,798
20	1	1.1	1	Preliminary	Yes	3	1.8	0.36	48,349	44.5	25,512	37,261	744	7,645	8,389
21	1	1.2	1	Preliminary	Yes	3	1.8	0.36	52,828	44.5	27,853	39,381	1,038	9,507	10,546
22	1	1.3	1	Preliminary	Yes	3	1.8	0.36	57,230	44.5	30,174	41,274	1,390	11,474	12,864
23	1	1.4	1	Preliminary	Yes	3	1.8	0.36	61,632	44.5	32,495	42,993	1,808	13,557	15,365
24	1	1.5	1	Preliminary	Yes	3	1.8	0.36	66,034	44.5	34,816	44,550	2,302	15,729	18,030
25	1	1	-7.2	Preliminary	Yes	3	1.8	0.36	44,023	37.3	35,654	33,999	172	6,917	7,089
26	1	1	-3.6	Preliminary	Yes	3	1.8	0.36	44,023	40.9	30,881	34,490	213	6,424	6,637
27	1	1	-1.8	Preliminary	Yes	3	1.8	0.36	44,023	42.7	27,276	34,744	332	6,211	6,542
28	1	1	1.8	Preliminary	Yes	3	1.8	0.36	44,023	46.3	18,801	35,359	672	5,803	6,475
29	1	1	3.6	Preliminary	Yes	3	1.8	0.36	44,023	48.1	14,457	35,695	887	5,609	6,496
30	1	1	7.2	Preliminary	Yes	3	1.8	0.36	44,023	51.7	7,592	36,365	1,323	5,176	6,499
RC	1	1	1	Revised	Yes	3	1.8	0.36	34,712	44.5	18,153	29,840	183	2,679	2,862

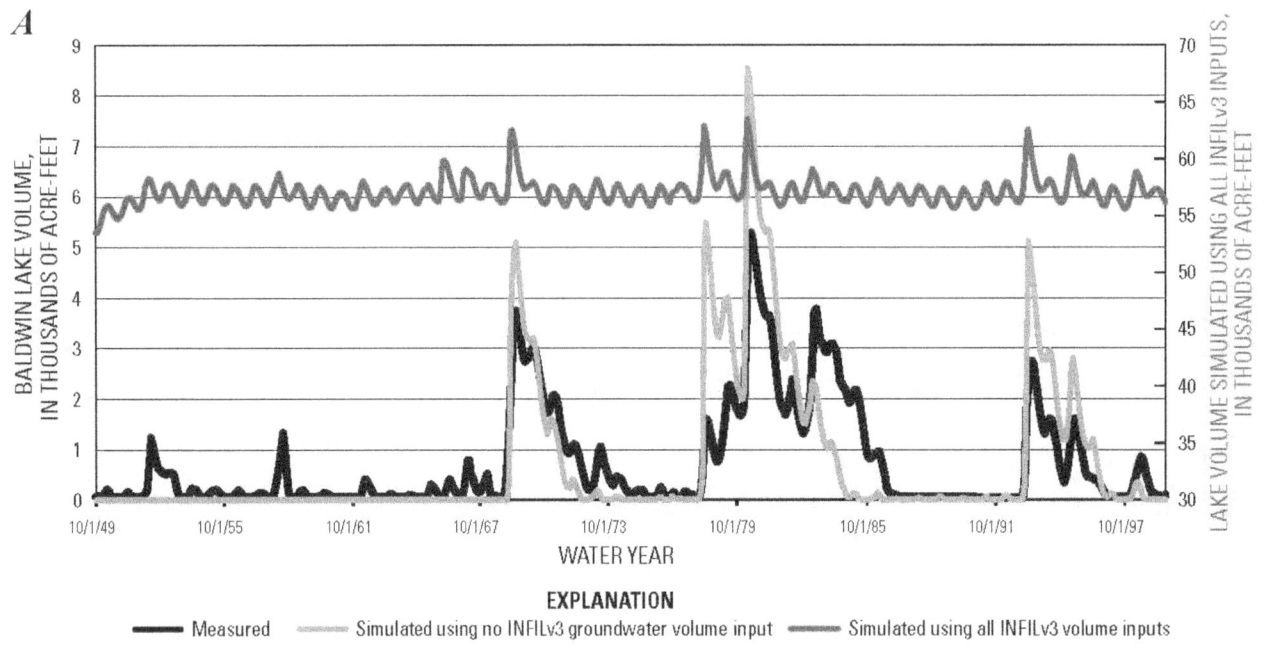

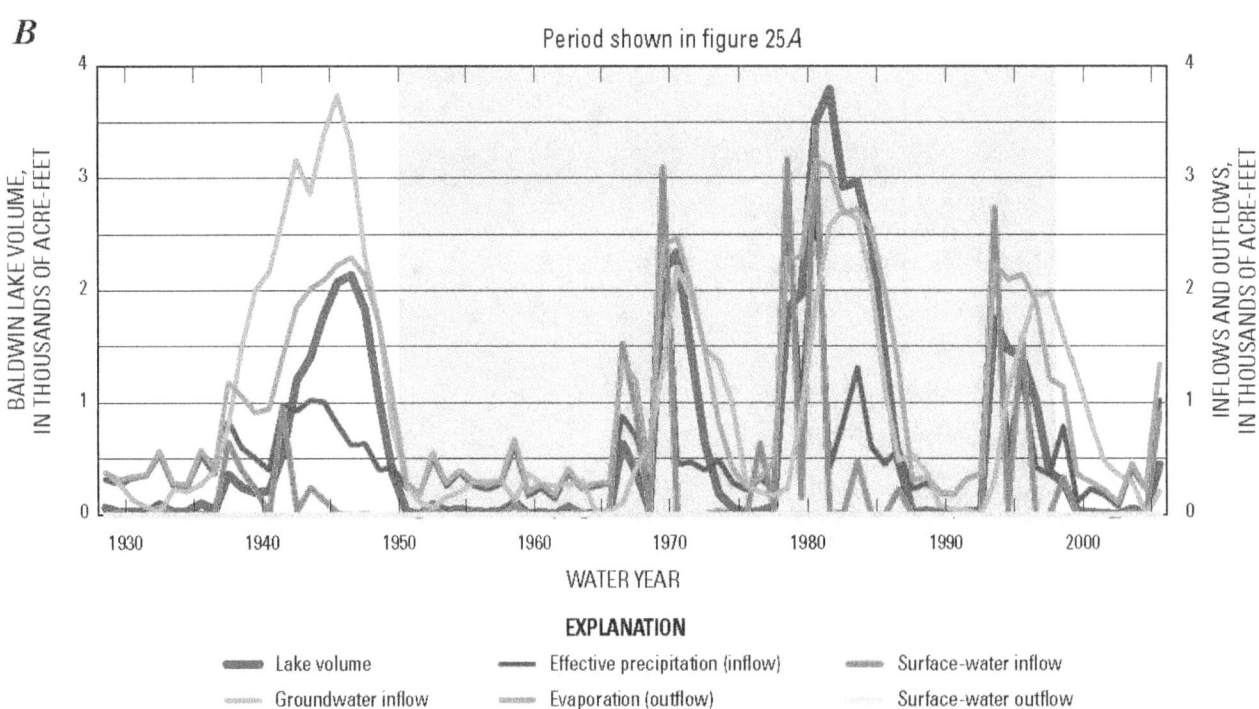

Figure 25. (A) Comparison of simulated and observed lake volumes for water-years 1950–99 and (B) simulated results for water years 1928–2005 for selected water-balance components of LAKE model of Baldwin Lake using the base-case INFILv3 model, Big Bear Valley, San Bernardino County, California.

To evaluate the effect of groundwater inflow on the simulated lake volume of Baldwin Lake, the groundwater inflow was set to zero for all time steps in the LAKE model. Eliminating groundwater inflow to the model resulted in a better match between measured and simulated lake volumes (*fig. 25A* and *table 10*), but this model configuration would require that groundwater outflow from the basin equal the total basin recharge of 5,994 acre-ft/yr, which is considered unlikely. The LAKE model results without groundwater inflows indicate that the timing of the surface-water inflows simulated by the INFILv3 base-case model were representative of hydrologic conditions for the Baldwin Lake drainage basin, but the absolute magnitude of the inflows was too high (*fig. 25B*).

Sensitivity Analysis of the Base-Case Model

The INFILv3 base-case model results indicate that the preliminary climate model adequately represents the distribution and quantity of precipitation in the Big Bear surface-water drainage basin but overestimates the quantity of precipitation in the Baldwin Lake surface-water drainage basin. As presented in the "Climate input" section of this report, the preliminary climate estimated with the NCDC stations and the PRISM data matched the two NCDC stations in the Big Bear surface-water drainage basin but overestimated the precipitation at the non-NCDC stations in the Baldwin Lake surface-water drainage basin (*fig. 20B*). Because the preliminary climate model does not adequately represent precipitation in the Baldwin Lake surface-water drainage basin, a sensitivity analysis was done to determine if variations in selected parameters and climate inputs could better match the measured volumes of Big Bear and Baldwin Lakes.

Twenty-nine alternative model configurations were included in the sensitivity analysis (*table 12*). Model comparisons were made using INFILv3 simulated evapotranspiration, recharge, and runoff; LAKE simulated lake volume and discharge; and the PAEE and NSME goodness-of-fit statistics for the simulation period, October 1, 1949, to September 30, 2005. The alternative models were identical to the INFILv3-LAKE base-case model configurations for Big Bear and Baldwin Lakes (*table 10*), with the exception of the differences in parameter values, model options, and climate inputs indicated in *table 12*.

Sensitivity in Vertical Hydraulic Conductivity

Models 1 through 12 (*table 13*) were used to analyze the sensitivity of simulation results to variations in parameters representing the perched zone, which are some of the more uncertain model inputs representing the physical characteristics of the surface-water drainage basins. Parameters that were adjusted in the analysis directly and indirectly affected simulated seepage, recharge, and runoff. Models 1 and 2 were used to compare the effect of differences in the seepage rate by using a 10-fold increase and decrease in the seepage factor. The seepage factor is a multiplier applied to each grid cell and was used to vary the upper horizontal (lateral) hydraulic conductivity of layer 6 uniformly over the model domain. Models 3 through 6 were used to test model sensitivity to variations in the vertical saturated hydraulic conductivities for layers 6 and 7, and also had the seepage component disabled by setting the seepage factor to zero. The vertical saturated hydraulic conductivities were adjusted uniformly for all grid cells according to the mapped rock types (*fig. 3, table 5*). Models 7 and 8 had higher seepage factors of 10 and 100, and also had higher values for the saturated hydraulic conductivities assigned to layers 6 and 7 (representing an increase in the permeability at the bottom of the SGWZ). Model 9 used a seepage factor of 1.0, but had the maximum range between the low and high saturated hydraulic conductivities assigned to the different geologic units. Model 10 used a modified spatial distribution of vertical hydraulic conductivities, which was based on rock type, with an overall increase in the basin-wide average hydraulic conductivities for layers 6 and 7. Models 11 and 12 were similar to model 10 in terms of the distribution of vertical hydraulic conductivities assigned to layers 6 and 7, but had thicker soils and higher seepage factors. Model 13 used a de-coupled runoff-model configuration (available as an option in INFILv3), thereby preventing runoff from infiltrating back into the root-zone during routing (all runoff is discharged). The de-coupled runoff configuration was used to evaluate the contribution of surface-water flow to recharge and streamflow.

None of the alternative models improved the model results in the Big Bear surface-water drainage basin compared to the base-case results, as indicated by the PAEE and NSME values (*table 12*). Models 1–5, 7, 8, and 10–13 provided satisfactory calibration results, indicating that varying the perched-zone properties did not result in failure to meet statistical goodness-of-fit criteria (PAEE ± 10, and NSME > 0.5); however, models 6 and 9 did not satisfy the calibration criteria. Model 6 simulated zero seepage and the highest vertical saturated hydraulic conductivity for layer 7. The lack of seepage, coupled with a high vertical hydraulic conductivity at the effective base of the model (layer 7), resulted in the highest simulated recharge of all the alternative models (9,790 acre-ft/yr), the highest total inflow to the lake (13,438 acre-ft/yr), and over-estimated the lake volumes (*table 12*). Model 9 allowed for seepage and had the greatest relative difference in the vertical saturated hydraulic conductivity between layers 6 and 7. The simulation of seepage with a low hydraulic conductivity at the effective base of the model (layer 7) resulted in the highest simulated evapotranspiration of the alternative models (35, 926 acre-ft/yr), the lowest recharge (1,283 acre-ft/yr), the lowest total inflow to the lake (5,813 acre-ft/yr), and a consistent underestimation of lake volumes (*table 12*).

Table 13. INFILv3 simulated average annual water-balance results determined by using the revised climate model for Big Bear and Baldwin Lakes surface-water drainage basins, San Bernardino County, California, water-years 1950–2005.

[Abbreviations: <, less than; ≥, greater or equal]

Subbasin	Inflow				Change in storage (water content)		Outflow			Recharge
	Total precipitaiton	Rainfall	Snowfall	Surface-water inflow	Root zone	Shallow groundwater zone	Sublimation	Evapotranspiration	Surface-water outflow	
Big Bear Lake surface-water drainage basin										
Big Bear Lake	796	458	338	3,907	3	0.2	34	639	3,989	37
Grout Creek	7,828	3,579	4,248	0	6	8.4	463	5,965	1,116	273
Village	3,872	1,985	1,887	0	13.7	5	164	3,304	132	255
Division	1,194	685	509	0	2.1	7.5	49	1,091	0	47
Mill Creek	9,326	4,300	5,026	0	13.8	5.9	452	6,252	2,173	429
Rathbone	8,307	3,862	4,446	0	28.2	18.1	395	6,962	30	887
North Shore	5,663	2,959	2,704	0	8	6.3	323	4,424	41	868
Gray's Landing	1,588	688	901	0	0	0.6	106	1,065	416	1
Average/total	38,574	18,516	20,058	0	74.9	52.1	1,986	29,702	3,989	2,798
Baldwin Lake surface-water drainage basin										
Van Dusen	7,564	3,360	4,204	0	19.9	6.7	468	6,198	110	763
West Baldwin	5,269	3,026	2,243	173	23.8	25.9	250	4,862	154	156
Erwin	16,279	6,973	9,307	0	42.6	34.8	901	13,665	63	1,586
East Baldwin	5,600	3,201	2,399	154	21.3	2.5	251	5,115	183	174
Average/total	34,712	16,559	18,153	0	107.6	69.8	1,871	29,840	183	2,679
Study area total	73,286	35,075	38,211	0	182.5	122	3,856	59,542	4,173	5,477

Table 13. INFILv3 simulated average annual water-balance results determined by using the revised climate model for Big Bear and Baldwin Lakes surface-water drainage basins, San Bernardino County, California, water-years 1950–2005.—Continued

[Abbreviations: <, less than; ≥, greater than or equal]

Subbasin	Basin-wide average water content (acre-feet)	Shallow groundwater zone — Average grid-cell water balance (inches per year)				Recharge indicators		
		Change in water content	Inflow — Net infiltration	Outflows — Seepage	Outflows — Recharge	Ratio of recharge to net outflow [1]	Ratio of seepage to recharge	Recharge as percent of precipitation
Big Bear Lake suface-water drainage basin								
Big Bear Lake	50	0.01	1.47	0.39	1.07	0.5	0.4	4.7
Grout Creek	1,026	0.02	1.34	0.55	0.76	0.2	0.7	3.5
Village	786	0.03	2.71	1.19	1.49	1.9	0.8	6.6
Division	125	0.06	1.78	1	0.72	130.4	1.4	4
Mill Creek	1,412	0.02	1.74	0.44	1.27	0.2	0.3	4.6
Rathbone	988	0.04	5.22	2.96	2.22	29.6	1.3	10.7
North Shore	374	0.01	5.99	2.92	3.06	21.2	1	15.3
Gray's Landing	96	0.01	0.1	0.07	0.02	0	3.5	0.1
Average/total	4,857	0.02	3.21	1.55	1.64	0.7	0.9	7.3
Baldwin Lake surface-water drainage basin								
Van Dusen	953	0.02	3.58	1.46	2.11	6.9	0.7	10.1
West Baldwin	190	0.03	1.06	0.54	0.49	–8	1.1	3
Erwin	2,049	0.03	3.8	1.8	1.98	25.3	0.9	9.7
East Baldwin	845	0.01	0.92	0.37	0.53	5.8	0.7	3.1
Average/total	4,037	0.02	2.75	1.25	1.48	14.6	0.8	7.7
Study area total	8,894	0.02	2.97	1.39	1.56	1.3	0.9	7.5

[1] Absolute values ≥1.0 identify recharge dominated subbasins; absolute values <1 identify runoff or outflow dominated subbasins. Negative values indicate subbasins where surface-water inflow exceeds surface-water outflow.

Similar to the base-case model, none of the alternative models 1–13 provided a good match to the Baldwin Lake record, as indicated by the goodness-of-fit statistics (*table 12*). The best results were obtained from models 9, 11, and 12, which had the smallest total inflows to Baldwin Lake and were the only models that did not result in spill-over of Baldwin Lake into the Big Bear Lake drainage basin. Overall, results for models 1 through 12 indicated that reasonable variations in the perched-zone properties did not improve the calibration results for Baldwin Lake while still maintaining a successful calibration for Big Bear Lake.

Sensitivity to Variations In Climate Input

Alternative models 14 through 29 were used to evaluate the sensitivity to daily precipitation and air temperature as defined by the preliminary-climate inputs. Models 14 through 23 used a multiplier for daily precipitation, with a minimum value of 0.5 and a maximum of 1.5 (*table 12*). Models 24 through 29 used an air temperature shift applied to maximum and minimum daily air temperature, with a minimum value of $-7.2°F$ and a maximum value of $7.2°F$.

As expected, the model results were sensitive to variations in the daily precipitation magnitude (models 14–23; *table 12*). Variations of more than plus or minus 10 percent precipitation (models 14–17, and models 20–23) resulted in PAEE and NSME values that did not satisfy the calibration criteria for the Big Bear Lake surface-water drainage basin (*table 12*). Decreasing the daily precipitation magnitude improved the model fit for the Baldwin Lake surface-water drainage basin (models 14–18); however, increasing the precipitation magnitude worsened the model fit (*table 12*). The best model fit for the Baldwin Lake surface-water drainage basin was achieved with a 20-percent reduction in precipitation (model 17; *table 12*).

Model sensitivity to variations in air temperature was less pronounced than to variations in precipitation for both Big Bear Lake and Baldwin Lake surface-water drainage basins (models 24–29; *table 12*). Lower air temperatures resulted in an increase in the percentage of precipitation occurring as snow, a decrease in evapotranspiration, a decrease in surface-water discharge, an increase in recharge, and higher average lake volumes. Calibration results for all six models included in the air-temperature analysis for Big Bear Lake surface-water drainage basin satisfied the goodness-of-fit criteria, although the base-case model still provided the best fit. All models resulted in unsatisfactory results in terms of the goodness-of-fit statistics, with the formation of a permanent lake and spill-over into the Bear Lake surface-water drainage basin.

Revised Climate Model

Because model sensitivity to climate input was found to be quite high, the climate model warranted revisions. As described in the "Model Input and Data Requirements" section, the preliminary-climate input was developed by using records from 35 selected NCDC climate stations (*fig. 20A, table 6*) and from monthly PRISM data. Average annual precipitation estimated by using the base-case model is higher than measured precipitation at the non-NCDC stations in the Big Bear study area that were omitted from the climate input (*fig. 20B*). The over-estimation of precipitation in the Baldwin Lake surface-water drainage basin is attributed to the local-scale rain-shadow effect on the leeward side of the Big Bear study area not being adequately represented by the regional-scale PRISM data.

To improve the match between measured and spatially-interpolated precipitation in the Baldwin Lake surface-water drainage basin, the PRISM monthly precipitation maps were revised, especially in the Baldwin Lake area. The revised PRISM monthly precipitation maps were modified by using the ratio of recorded average monthly precipitation to PRISM average monthly precipitation at 84 NCDC, RAWS, CIMIS, and SBC stations having at least 10 years of record for a given month (*table 6*). The calculated ratios were spatially interpolated for the study area by using GIS and the inverse-distance squared method to generate a map of ratios for each month (*fig. 26A*). The ratio maps were then multiplied by the original PRISM maps to produce a revised PRISM map for each month. The revised PRISM map for the month of January provides an example of the revised spatial distribution of average precipitation with local precipitation data (*fig. 26B*). Then the revised PRISM monthly precipitation maps were used in the spatial-interpolation model to estimate daily precipitation for all INFILv3 grid cells, as described previously.

The spatial distribution of average annual precipitation estimated using the revised climate input (*fig. 27*) indicated decreased precipitation compared to average annual precipitation estimated using the preliminary-climate input (*fig. 20B*). The decreased precipitation was most pronounced in the Baldwin Lake surface-water drainage basin where the modeled precipitation was reduced from 22 to approximately 18 in/yr. This provided a better overall match to measured average annual precipitation in the Big Bear study area. The highest average precipitation (30 in/yr and greater) was estimated for the western-most part of the Big Bear study area (in the vicinity of Bear Valley dam) and the lowest average precipitation (18–20 in/yr) was estimated along the northwestern boundary of the East Baldwin subbasin (*fig. 27*).

The spatial distribution of snow estimated using the revised climate input was matched to the DEM contours, reflecting the linear monthly air temperature – altitude regression models used to spatially-distribute maximum and minimum daily air temperature (*fig. 28*). Less than 40 percent of the estimated precipitation occurred as snow for the lower altitudes within the study area, corresponding to estimated average air temperatures of 46 to 47°F (*fig. 21*). Locations where 80 percent or more of the estimated precipitation occurred as snow corresponded to the higher altitudes in the Big Bear study area and to locations where the estimated average air temperatures were approximately 38°F and lower.

A

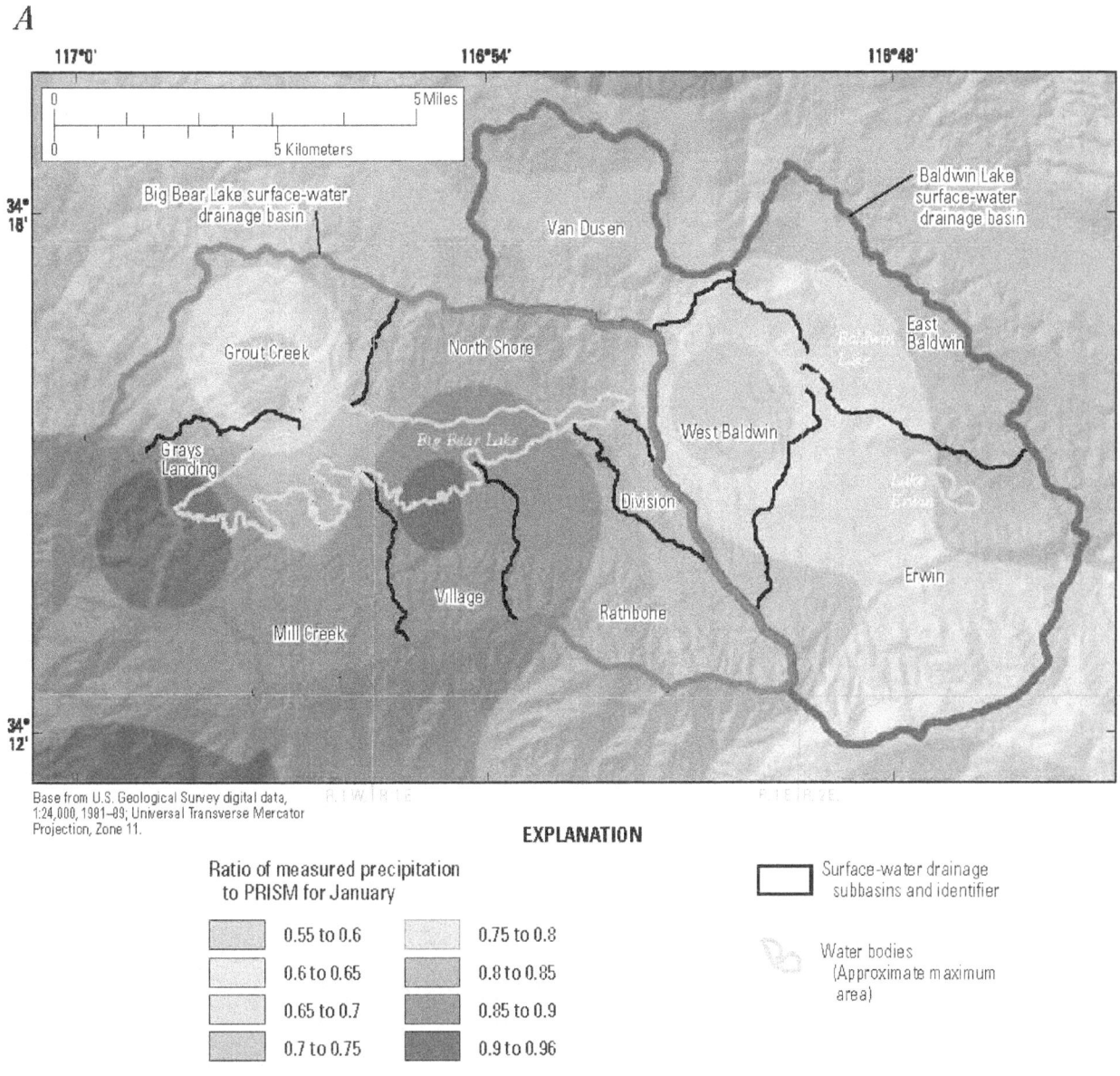

Base from U.S. Geological Survey digital data,
1:24,000, 1981–89; Universal Transverse Mercator
Projection, Zone 11.

EXPLANATION

Ratio of measured precipitation
to PRISM for January

0.55 to 0.6	0.75 to 0.8
0.6 to 0.65	0.8 to 0.85
0.65 to 0.7	0.85 to 0.9
0.7 to 0.75	0.9 to 0.96

Surface-water drainage
subbasins and identifier

Water bodies
(Approximate maximum
area)

Figure 26. (*A*) ratios of measured precipitation station data to PRISM data interpolated to the study area for the month of January and (*B*) the modified PRISM average monthly precipitation map for the month of January for the Big Bear area, San Bernardino County, California.

Revised Climate-Model Calibration

The base-case INFILv3 model was used to simulate the daily water balance for the Big Bear area with the revised climate input, referred to as the "revised climate model" in this report. With the exception of the revised climate input, all other INFILv3 model parameters were the same as for the base-case model, and model calibration consisted of adjusting only LAKE model parameters for both Big Bear and Baldwin Lakes. The INFILv3 revised climate model simulated-average annual precipitation (rain and snow) for water years 1950–2005 in the Big Bear study area was about was about

73,000 acre-ft/yr, or about a 17 percent reduction (about 14,900 acre-ft/yr) compared to the base-case model (*tables 10, 13*). The simulated-average annual precipitation was about 38,600 acre-ft/yr for the Big Bear Lake surface-water drainage basin and about 34,700 acre-ft/yr for the Baldwin Lake surface-water drainage basin. The simulated-average annual recharge to the Big Bear Lake surface-water-drainage basin was about 2,800 acre-ft/yr (about 7.3 percent of precipitation), and the average annual surface-water outflow was about 3,990 acre-ft/yr (about 10.3 percent of precipitation; *table 14*). The simulated-average annual recharge to the Baldwin Lake

B

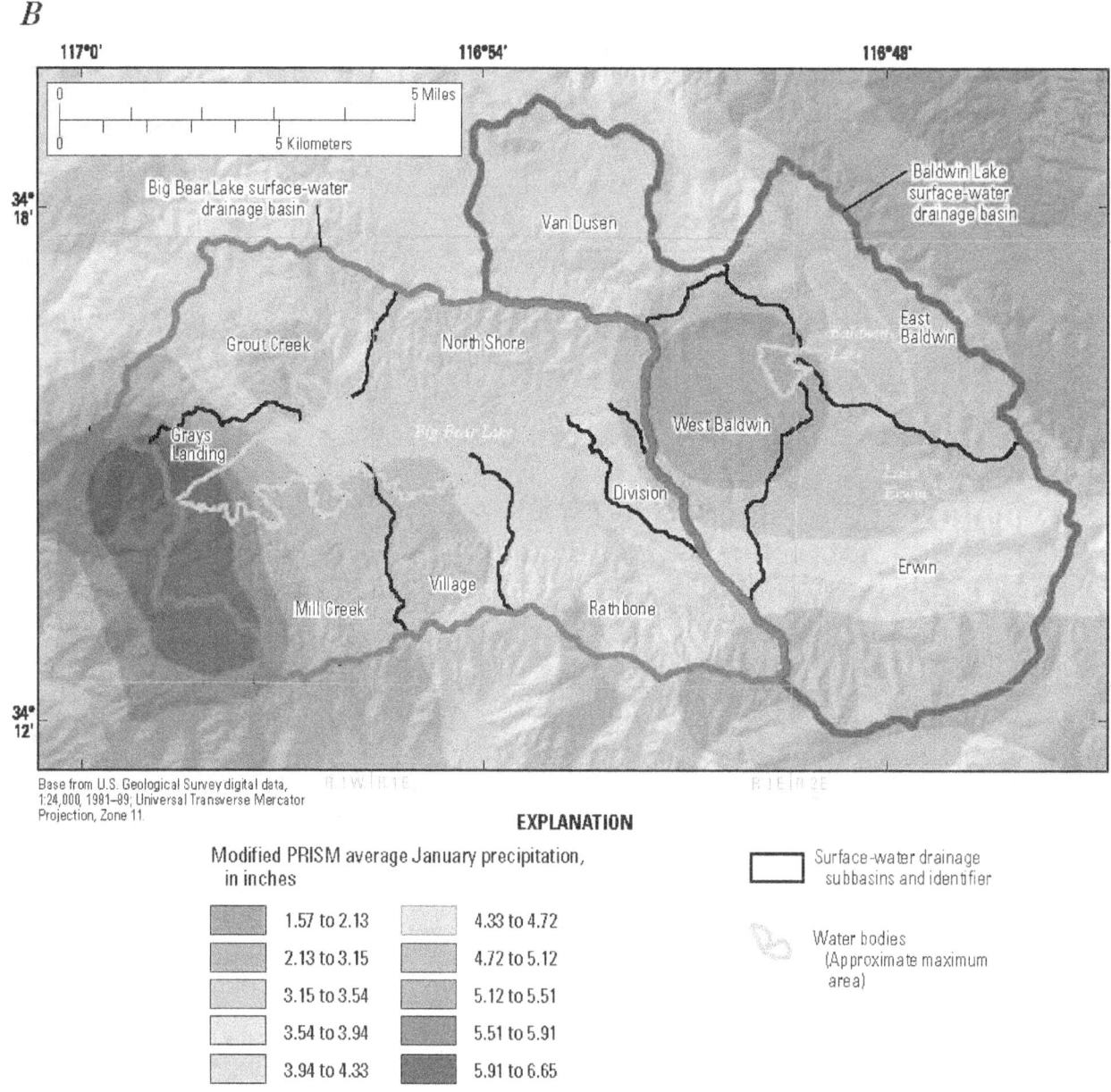

EXPLANATION

Modified PRISM average January precipitation, in inches

1.57 to 2.13	4.33 to 4.72
2.13 to 3.15	4.72 to 5.12
3.15 to 3.54	5.12 to 5.51
3.54 to 3.94	5.51 to 5.91
3.94 to 4.33	5.91 to 6.65

☐ Surface-water drainage subbasins and identifier

Water bodies (Approximate maximum area)

Base from U.S. Geological Survey digital data, 1:24,000, 1981–89; Universal Transverse Mercator Projection, Zone 11.

Figure 26.   Continued

surface-water-drainage basin was about 2,680 acre-ft/yr (about 7.7 percent of precipitation), and the average annual surface-water outflow was about 183 acre-ft/yr (about 0.5 percent of precipitation; *table 13*). Comparison of results between the base-case and revised climate models illustrates the non-linear relation between simulated precipitation and recharge for the study area. Reducing the base-case average annual precipitation by about 17 percent resulted in about a 45-percent reduction in simulated recharge in the revised climate model (*tables 10* and *13*).

Revised Climate-Model Calibration for Big Bear Lake

The INFILv3 revised climate model simulated surface-water outflow and recharge for the Big Bear Lake surface-water drainage basin were used as inflows to the LAKE model for Big Bear Lake. The best fit to estimated lake volumes was achieved by modifying the following LAKE parameters calibrated for the base-case model: (1) increasing the steady-state groundwater inflow fraction from 0.1 to 0.7, (2) decreasing the transient groundwater-inflow fraction from 0.9 to 0.3, (3) increasing the transient recharge averaging period from 2 to

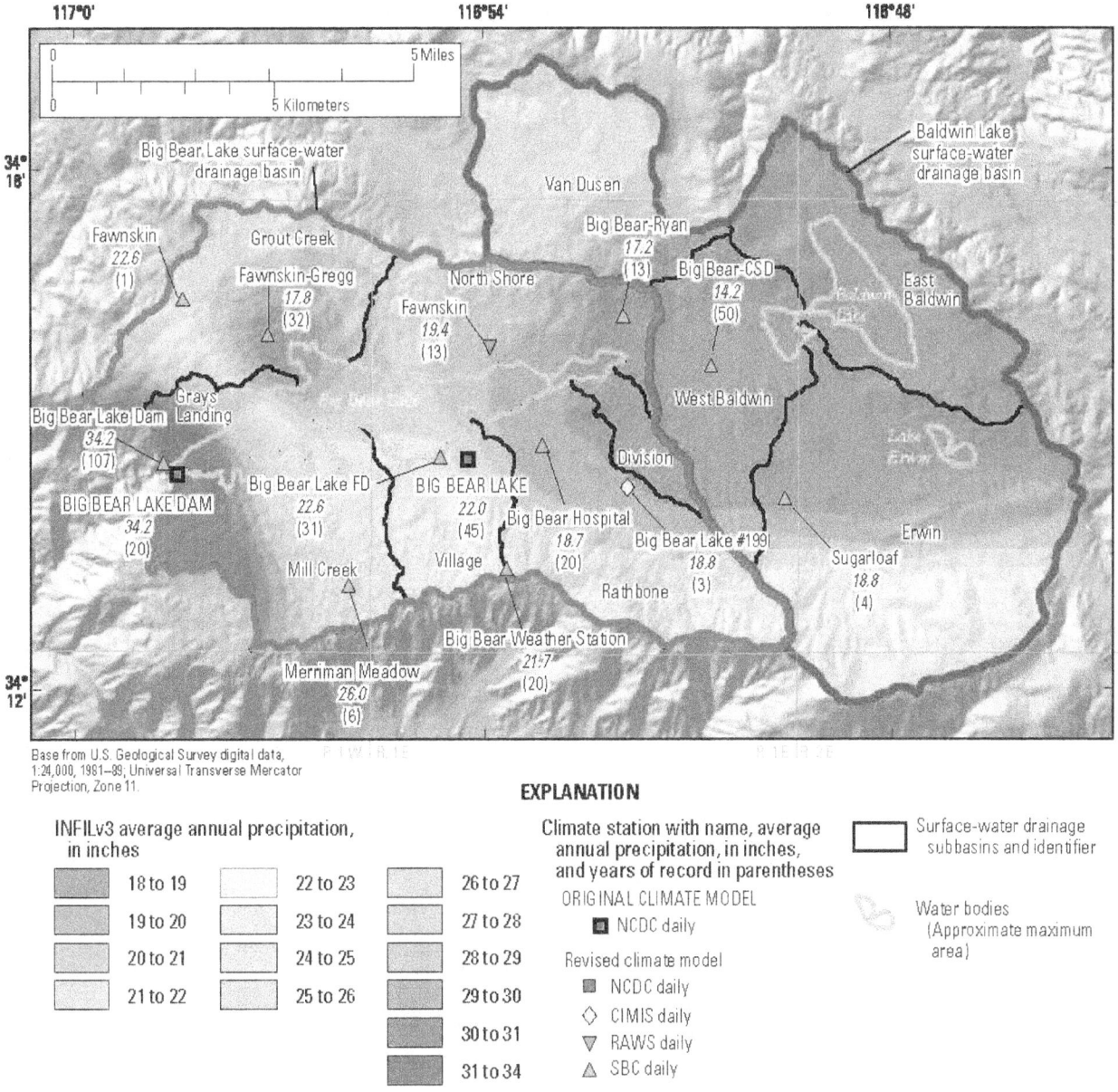

Figure 27. Comparison of average annual precipitation simulated by using the revised climate INFILv3 model and measured precipitation from all meteorological stations (NCDC and supplemental stations) in the Big Bear area, San Bernardino County, California.

3 years, (4) increasing the minimum fraction of groundwater inflow to the wetted area from 0.25 to 1.0, and (5) increasing the Hortonian runoff fraction from 0.6 to 0.8.

A comparison of the simulated and estimated lake volumes for the calibration period (calendar years 1985–2005) indicates a difference of less than 5,000 acre-ft/yr over the period of record (*fig. 29A*); however the difference is greater compared to the base-case model (*fig. 24A*). The average simulated Big Bear Lake volume was 52,135 acre-ft compared to the estimated initial lake volume of 56,049 (*table 14*). The

goodness-of-fit statistics were well within the limits for a successful calibration (*table 14*); PAEE was –6.31 percent, NSME was 0.80, and the r^2 was 0.94 (*table 14*). If none of the simulated INFILv3 recharge was used in the LAKE model, it under-estimated lake volumes, resulting in unsatisfactory goodness-of-fit statistics (*fig. 29A, table 14*), which demonstrates the importance of groundwater discharge to the Big Bear Lake water budget.

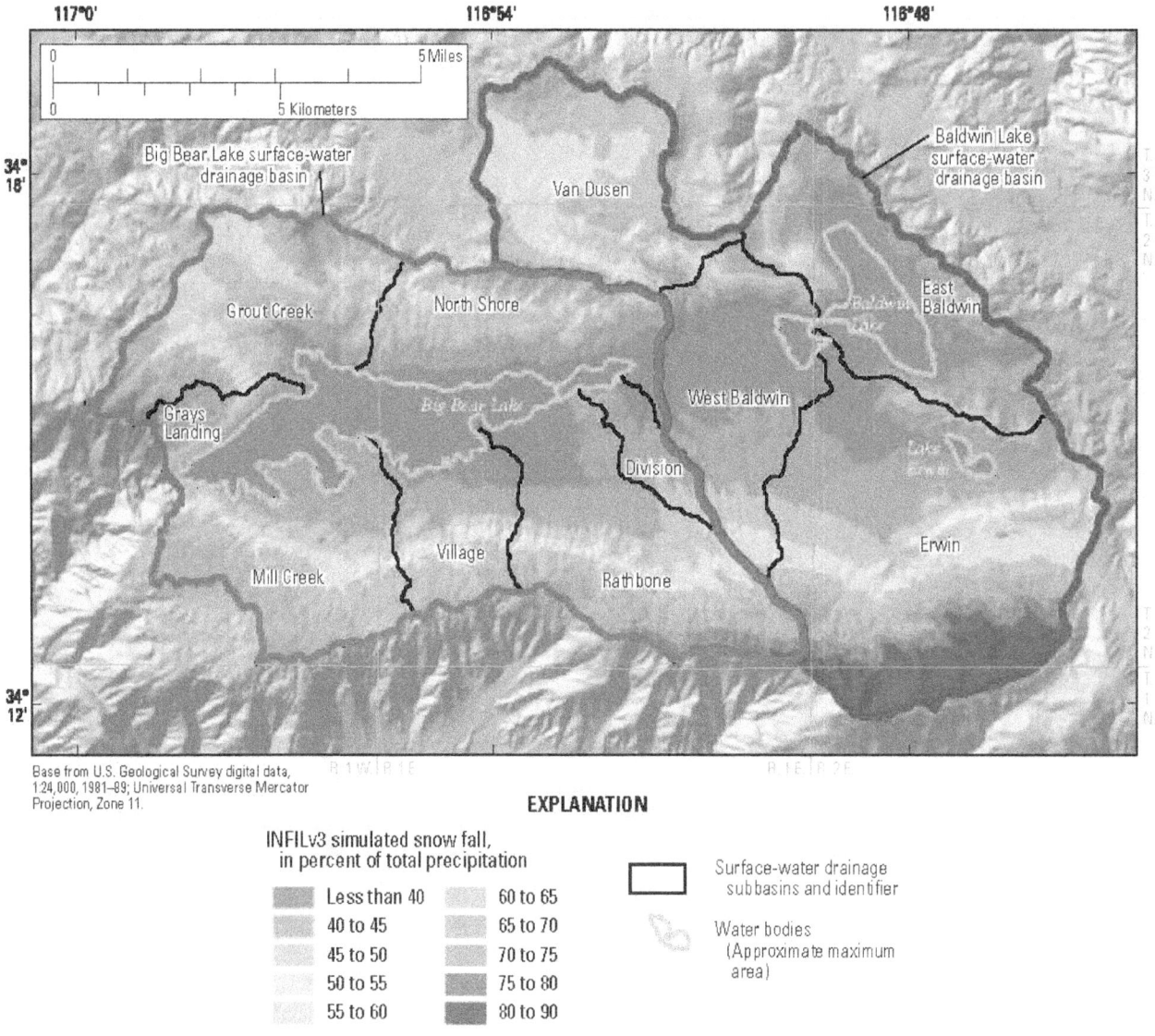

Base from U.S. Geological Survey digital data,
1:24,000, 1981–89; Universal Transverse Mercator
Projection, Zone 11.

EXPLANATION

INFILv3 simulated snow fall,
in percent of total precipitation

Less than 40		60 to 65	
40 to 45		65 to 70	
45 to 50		70 to 75	
50 to 55		75 to 80	
55 to 60		80 to 90	

Surface-water drainage
sub basins and identifier

Water bodies
(Approximate maximum
area)

Figure 28. Revised climate INFILv3 model simulated snowfall, as percent of total precipitation for the Big Bear area, San Bernardino County, California.

The revised climate-model annual simulation results for water-years 1928–2005 indicated a high degree of year-to-year variability in inflows, outflows, and changes in lake volume for Big Bear Lake (*fig. 29B*). Unlike results obtained using the base-case model, simulated annual discharges at Big Bear Lake dam were less than the simulated annual surface-water inflows to the lake. Annual discharges greater than 8,000 acre-ft occurred for only two water years, 1969 and 1980. All annual discharges after water year 1980 were less than 3,000 acre-ft. For most water years during1928–2005, total annual discharge from the lake was less than 1,000 acre-ft. Annual variability in simulated groundwater discharge to the lake was lower than the base-case model, with annual inflows less than 4,000 acre-ft for most years. Similar to the base-case model results, effective precipitation over the lake area was

an important inflow component to the Big Bear Lake water balance.

Revised Climate-Model Calibration for Baldwin Lake.

The INFILv3 revised climate model simulated surface-water outflow and recharge for the Baldwin Lake surface-water drainage basin were used as inflows to the LAKE model for Baldwin Lake. The best fit to measured lake volumes was achieved by using the following LAKE parameters: (1) steady-state groundwater-inflow fraction set to 0.75, (2) transient groundwater-inflow fraction set to 0.25, (3) transient recharge averaging period set to 5 years, (4) minimum fraction of groundwater inflow to the wetted area set to 0.0, and (5) the Hortonian runoff fraction set to 0.2.

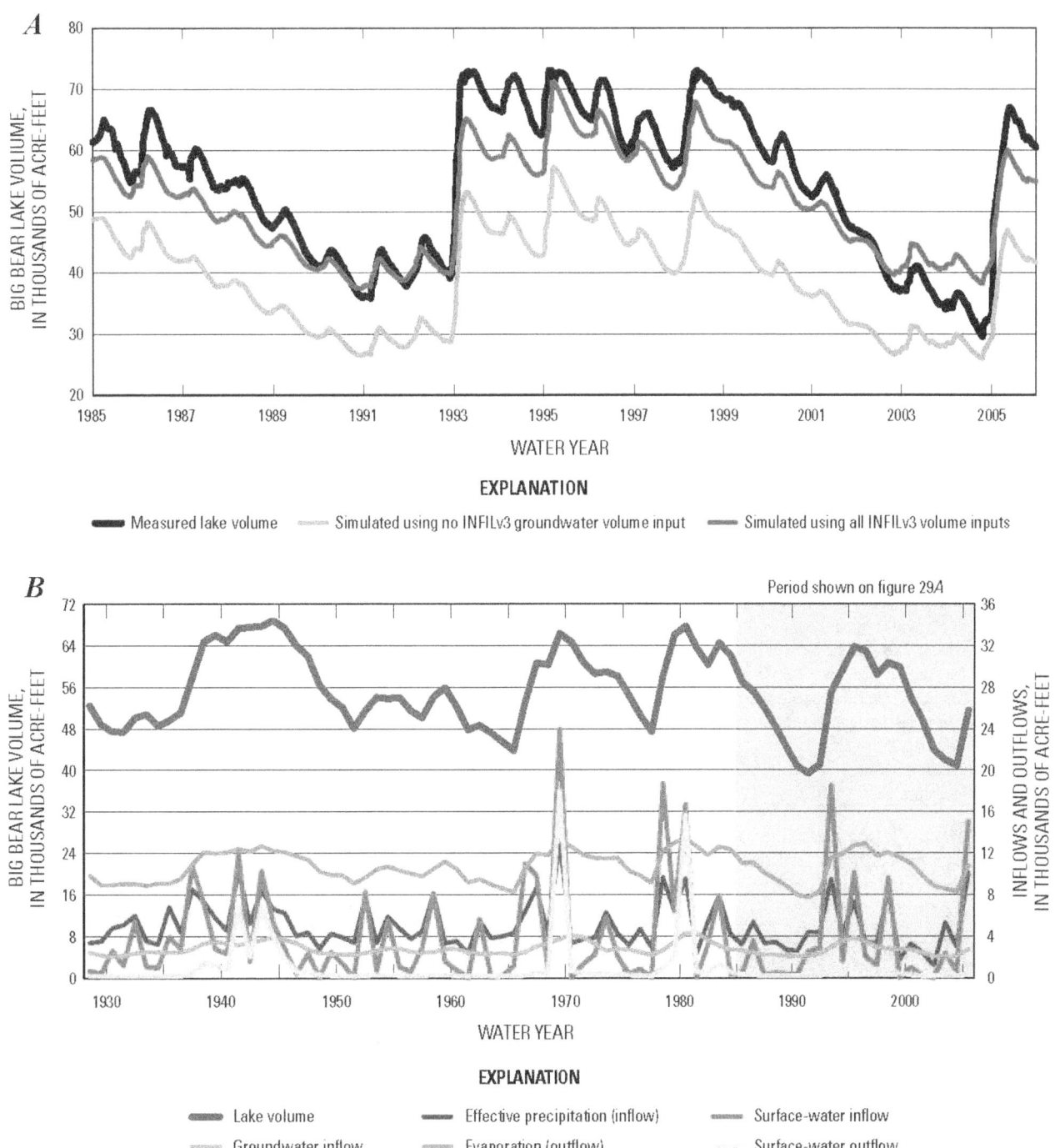

Figure 29. (A) Comparison of simulated and observed lake volumes for calendar years 1985–2005 and (B) simulated results for water-years 1930–2005 for selected water-balance components of the LAKE model of Big Bear Lake by using the revised INFILv3 climate model, Big Bear Valley, San Bernardino County, California.

Table 14.   LAKE model results, calibration statistics, and model parameters using the revised INFILv3 climate model, Big Bear and Baldwin Lakes, San Bernardino County, California.

[**Abbreviations:** ac-ft, acre-feet; –, no data available]

Parameter	Units	INFILv3 revised climate model, Big Bear Lake		INFILv3 revised climate model, Baldwin Lake	
Revised-climate model results and calibration statistics					
INFILv3 revised-climate model inflow volumes	–	No reduction in INFILv3 inflows	No groundwater inflow	No reduction in INFILv3 inflows	No groundwater inflow
Simulated mean lake level	Feet	64.7	59.5	2.5	0.8
Simulated maximum lake level	Feet	71.6	66.7	11.2	8.5
Simulated minimum lake level	Feet	58.6	53	0	0
Simulated mean lake volume	Ac-ft	52,513	39,793	737	203
Simulated maximum lake volume	Ac-ft	71,171	57,196	4,882	3,167
Simulated minimum lake volume	Ac-ft	37,426	26,032	0	0
Percent average estimation error	Percent	–6.31	–29	0.73	–72.24
Nash-Sutcliffe model efficiency	–	0.8	–0.97	0.79	0.03
R-squared	–	0.94	0.94	0.8	0.3
LAKE model parameters (revised climate model)					
Fraction of INFILv3 precipitation	Decimal	1	1	1	1
Fraction of INFILv3 surface-water inflow	Decimal	1	1	1	1
Fraction of INFILv3 groundwater inflow	Decimal	1	0	1	0
Steady-state groundwater-inflow fraction	Decimal	0.7	0.7	0.75	0.75
Transient groundwater-inflow fraction	Decimal	0.3	0.3	0.25	0.25
Transient recharge averaging period	Years	3	3	5	5
Upstream groundwater-discharge area	Acres	0	0	192.23	192.23
Minimum fraction surface-water inflow to wetted area	Decimal	1	1	0.3	0.3
Minimum fraction groundwater inflow to wetted area	Decimal	1	1	0	0
Maximum soil-zone storage capacity	Inches	20	20	24	24
Initial soil-zone storage capacity	Inches	5	5	5	5
Hortonian runoff fraction	Decimal	0.8	0.8	0.2	0.2
Estimated lake volume	Ac-ft	56,049	56,049	731	731

A comparison of the revised climate model simulated and estimated lake volumes for the calibration period (water-years 1950–1999) indicated a difference of less than 1,000 acre-ft/yr for the period of record (*fig. 30A*), unlike the comparison for the base-case model (*fig. 25A*). The average simulated Baldwin Lake volume was 737 acre-ft/yr compared to the measured lake volume of 731 (*table 14*). The goodness-of-fit statistics were well within the limits for a successful calibration, with a PAEE of 0.73 percent, a NSME of 0.79, and an r^2 of 0.80 (*table 14*). Similar to the Big Bear Lake results, if none of the simulated INFILv3 recharge for the Baldwin Lake surface-water drainage basin was used in the LAKE model, the model under-estimated lake volumes, which produced unsatisfactory goodness-of-fit statistics (*fig. 30A, table 14*).

The revised climate-model annual simulation results for water-years 1928–2005 showed a decrease in the frequency of water years having significant surface-water discharge to the lake, along with a decrease in the magnitude of annual inflows, compared to the base-case model (*figs. 25B, 30B*). Annual groundwater discharge estimated by using the revised climate model was between 1,000 and 3,000 acre-ft throughout the

simulation period, and was less variable on a year-to-year basis compared to the base-case model because of the much higher percentage of the total flow simulated as a steady-state discharge component. Groundwater discharge was the primary inflow component to the lake water balance for most water years.

Simulation Results

The INFILv3 revised climate-model results were used in the LAKE model to evaluate the inflow and outflows from Big Bear and Baldwin Lakes and the quantity and spatial distribution of evapotranspiration, surface-water runoff, water content of the shallow groundwater zone, seepage, and recharge for the entire Big Bear and Baldwin Lakes surface-water drainage basin for water-years 1950–2005. As previously discussed (see "Climate input" section), water-years 1950–2005 had the greatest number of climate stations with data, making this period the most appropriate for simulating water balance (including recharge) in the Big Bear Valley study area.

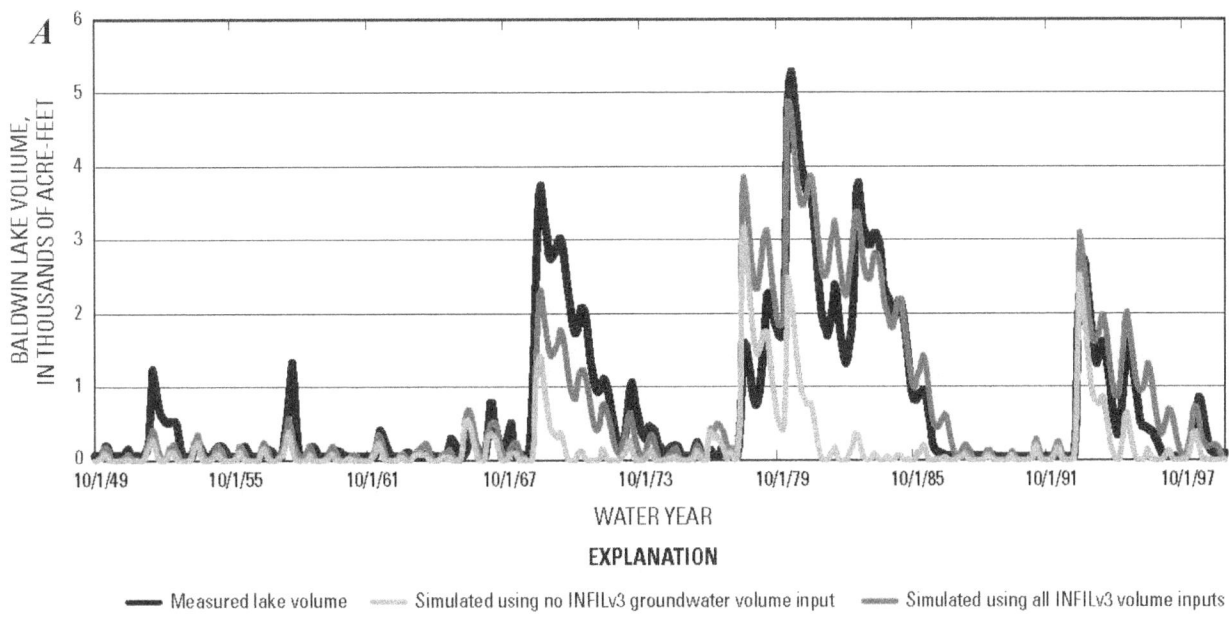

EXPLANATION

━━━ Measured lake volume   ━━━ Simulated using no INFILv3 groundwater volume input   ━━━ Simulated using all INFILv3 volume inputs

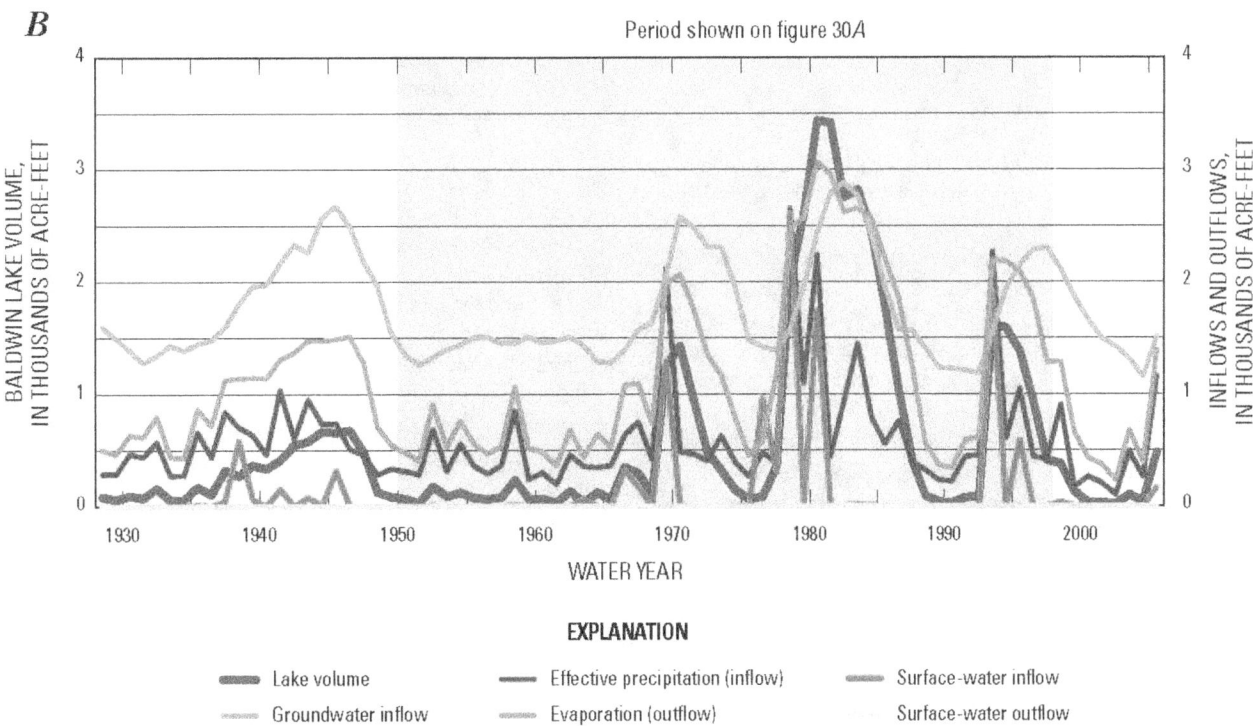

EXPLANATION

━━━ Lake volume   ━━━ Effective precipitation (inflow)   ━━━ Surface-water inflow
━━━ Groundwater inflow   ━━━ Evaporation (outflow)   ━━━ Surface-water outflow

Figure 30. (*A*) Comparison of simulated and observed lake volumes for water years 1950–1999 and (*B*) simulated results for water years 1928–2005 for selected water-balance components of the LAKE model of Baldwin Lake using the INFILv3 revised climate model, Big Bear Valley, San Bernardino County, California.

LAKE Model Results

Simulated average annual LAKE model results for Big Bear and Baldwin Lakes water-years 1950–2005 are presented on *table 15*. Precipitation for the total area of Big Bear Lake (about 5,330 acre-ft/yr) was the largest source of inflow, followed by surface-water inflow (about 3,990 acre-ft/yr) and groundwater discharge (about 2,800 acre-ft/yr). Evapotranspiration from the total lake area was by far the largest component of outflow from the lake (about 10,910 acre-ft/yr), followed by surface-water outflow (about 990 acre-ft/yr) and sublimation from the dry lakebed (about 155 acre-ft/yr). Over the simulation period, there was about a 65 acre-ft/yr loss of storage from the lake.

Groundwater discharge was the largest source of inflow to Baldwin Lake (about 1,720 acre-ft/yr), followed by precipitation (about 1,440 acre-ft/yr) and surface-water inflow (about 185 acre-ft/yr) (*table 15*). Similar to Big Bear Lake, evapotranspiration from the total lake area was by far the largest component of outflow from Baldwin Lake (about 3,220 acre-ft/yr), with sublimation from the dry lakebed (about 125 acre-ft/yr) being the only other simulated outflow from the lake. Over the simulation period, there was almost no change in storage from the lake.

Evapotranspiration

Evapotranspiration is the largest simulated component of outflow from the Big Bear study area, averaging about 60,000 acre-ft/yr during water-years 1950–2005 (*table 13*). About 80 percent of the simulated total precipitation (about 73,290 acre-ft/yr) is lost to evapotranspiration. The spatial distribution of simulated average annual evapotranspiration ranged from less than 12 to 43 in/yr (*fig. 31*). The lowest values were from north-facing slopes, where potential evapotranspiration was at a minimum. In addition, areas of low simulated evapotranspiration corresponded to locations that had thin soils, low vegetation density, and relatively permeable bedrock underlying the root zone. The greatest values of evapotranspiration were found in areas having deep root zones (thick soils underlain by unconsolidated or alluvial deposits), especially along stream channels receiving frequent runoff, and areas having higher root densities, such as forested areas.

Surface-Water Runoff

The average annual simulated surface-water outflow from the Big Bear study area for water-years 1950–2005 was about 4,170 acre-ft/yr, with about 3,990 acre-ft/yr in the Big Bear surface-water drainage basin and about 180 acre-ft/yr in the Baldwin surface-water drainage basin (*table 13*). The spatial distribution of simulated average annual surface-water runoff was computed as the average depth in inches per year (*fig. 32*). The spatial distribution of simulated average annual runoff indicated a wide range in values for the main drainages. The accumulation of overland flow in the smaller first-order channels resulted in relatively low average runoff depths, ranging from 50 to 200 in/yr for all subbasins and model units. Intermediate runoff depths ranged from 200 to 2,000 in/yr for sections of the main channels in most of the drainages in the study area. Runoff depths greater than 2,000 in/yr occurred in the main channels of the larger subbasins: Erwin and Van Dusen in the Baldwin Lake drainage basin, and all subbasins, except Division, in the Big Bear Lake drainage basin.

For many subbasins, the maximum simulated runoff depth did not occur at the outflow grid cell, but rather in the middle or upper sections of the subbasin. These subbasins included North Shore, Village, and Rathbone in the Big Bear Lake surface-water drainage basin, and Erwin, West Baldwin, and East Baldwin in the Baldwin Lake surface-water drainage basin. Simulated runoff in the main channels of these subbasins decreased further downstream, toward Big Bear and Baldwin Lakes, because part of the runoff infiltrated into the channel bed in areas with thick soils and permeable bedrock.

Water Content of the Shallow Groundwater Zone

The total simulated basin-wide average water content of the shallow groundwater zone was about 8,900 acre-ft (about 2.3 in.) for the simulation period (water-years 1950–2005), or about 0.2 percent of the total simulated precipitation (*table 13*). The spatial distribution of the simulated average annual water content of the shallow groundwater zone (*fig. 33*) indicated a variable pattern that depends on topography, net infiltration through the root zone, and bedrock and soil saturated hydraulic conductivity. Water content of less than 1 in. occurred over relatively wide areas, including most of the inter-channel locations in the permeable unconsolidated deposits and the relatively permeable, consolidated deposits. Water content of more than 20 in. occurred where permeable bedrock was downstream of areas generating high surface-water and seepage inflows (areas characterized by thin soils, impermeable bedrock, low vegetation density, and low PET). These areas also were characterized by thick soils overlying unconsolidated deposits caused by the high surface-water inflows.

Lateral Seepage

The spatial distribution of simulated 1950–2005 average annual lateral seepage (*fig. 34*) indicated a dependency on topography, the permeability (hydraulic conductivity) of bedrock or unconsolidated deposits underlying the root zone, and net infiltration from the root zone. High seepage values were simulated for locations with high net infiltration, permeable rock types, and steep slopes. These combined effects resulted in seepage values exceeding 50 in/yr for the higher altitudes in the Erwin, Rathbone, Van Dusen, and North Shore subbasins (*fig. 34*). Low seepage values were simulated for locations with thin soils underlain by consolidated bedrock having low permeability (granitic and metamorphic rock types).

Table 15. Results for LAKE model simulations for water-years 1950–2005, from the INFILv3 revised climate model for Big Bear Lake and Baldwin Lake, San Bernardino County, California.

[Abbreviations: ac-ft, acre-feet; GW, groundwater; PET, potential evapotranspiration; –, not applicable]

Parameter	Units	Water-balance component	Revised climate model: Big Bear Lake				Revised climate model: Baldwin Lake			
			Average annual	Average daily	Maximum daily	Minimum daily	Average annual	Average daily	Maximum daily	Minimum daily
Total upstream land area	Acres	–	20,487	20,487	20,487	20,487	21,741	21,741	21,741	21,741
Groundwater-discharge area	Acres	–	0	0	0	0	192	192	192	192
Total lake area	Acres	–	2,854	2,854	2,854	2,854	1,079	1,079	1,079	1,079
Dry-lakebed area	Acres	–	742	742	1,391	0	835	835	1,052	378
Water area	Acres	–	2,112	2,112	2,854	1,463	243	243	701	26
Total upstream land area (includes groundwater-discharge area)										
Total net-infiltration to subbasin	Ac-ft	Inflow	2,798.20	7.7	255.7	0.3	2,678.70	7.3	396.1	0.2
Total groundwater discharge in subbasin	Ac-ft	–	2,798.20	7.7	13.5	5.6	2,678.60	7.3	10.6	5.8
Surface-water runoff	Ac-ft	Inflow	3,989.50	10.9	7,377.10	0	183.4	0.5	1,025.10	0
Deep GW recharge to region	Ac-ft	Outflow	0	0	0	0	0	0	–	–
Evapotranspiration from groundwater	Ac-ft	Outflow	0	0	0	0	958.5	2.6	5.1	0.3
Groundwater discharge to lake area	Ac-ft	Outflow	2,798.20	7.7	13.5	5.6	1,720.20	4.7	10	1
Surface-water discharge to lake area	Ac-ft	Outflow	3,989.50	10.9	7,377.10	0	183.4	0.5	1,025.10	0
Total lake area (dry-lakebed and water areas)										
PET for total lake area	Ac-ft	–	14,334.70	39.2	75.4	2.7	5,378.00	14.7	28.5	1.9
Precipitation on total lake area	Ac-ft	Inflow	5,329.20	14.6	2,087.80	0	1,442.40	3.9	453.6	0
Surface-water flow to lake area	Ac-ft	Inflow	3,989.50	10.9	7,377.10	0	183.4	0.5	1,025.10	0
Groundwater flow to lake area	Ac-ft	Inflow	2,798.20	7.7	13.5	5.6	1,720.20	4.7	10	1
Change in water storage	Ac-ft	Storage	64.7	0.2	–	–	4	0	–	–
Sublimation from dry lakebed	Ac-ft	Outflow	153.9	0.4	8.3	0	124.2	0.3	6.3	0
Evapotranspiration	Ac-ft	Outflow	10,909.30	29.9	–	–	3,217.90	8.8	–	–
Surface-water outflow	Ac-ft	Outflow	989	2.7	2,641.40	0.6	0	0	0	0
Dry-lakebed area										
Precipitation on dry lakebed	Ac-ft	–	1,368.10	3.7	666.4	0	1,074.00	2.9	365	0
Rain on dry lakebed	Ac-ft	Inflow	781.6	2.1	666.4	0	675	1.8	365	0
Snow on dry lakebed	Ac-ft	Inflow	586.5	1.6	397.2	0	399	1.1	223.4	0
Surface-water inflow to dry lakebed	Ac-ft	Inflow	0	0	0	0	83.9	0.2	462.2	0
Groundwater inflow to dry lakebed	Ac-ft	Inflow	0	0	0	0	1,275.10	3.5	7.2	0.9
Change in soil-zone storage	Ac-ft	Storage	0	0	125.9	-134.4	0	0	598.6	-216.8
Sublimation from dry lakebed	Ac-ft	Outflow	153.9	0.4	8.3	0	124.2	0.3	6.3	0
Evapotranspiration from dry lakebed	Ac-ft	Outflow	240.1	0.7	26.3	0	2,049.40	5.6	25.7	0.9
Runoff from dry lakebed to lake	Ac-ft	Outflow	974.1	2.7	533.1	0	259.3	0.7	584.8	0

Table 15. Results for LAKE model simulations for water-years 1950–2005, from the INFILv3 revised climate model for Big Bear Lake and Baldwin Lake, San Bernardino County, California.—Continued

[Abbreviations: ac-ft, acre-feet; GW, groundwater; PET, potential evapotranspiration; —, not applicable]

Parameter	Units	Water-balance component	Revised climate model: Big Bear Lake				Revised climate model: Baldwin Lake			
			Average annual	Average daily	Maximum daily	Minimum daily	Average annual	Average daily	Maximum daily	Minimum daily
			Water area							
Water volume	Ac-ft	—	54,021.30	54,021.30	78,663.40	37,426.30	668.8	668.8	4,963.50	0
Precipitation on water area	Ac-ft	Inflow	3,961.10	10.8	1,745.40	0	368.5	1	142.6	0
Runoff from dry lakebed to lake	Ac-ft	Inflow	974.1	2.7	533.1	0	259.3	0.7	584.8	0
Surface-water inflow	Ac-ft	Inflow	3,989.50	10.9	7,377.10	0	99.6	0.3	587.1	0
Groundwater inflow	Ac-ft	Inflow	2,798.20	7.7	13.5	5.6	445.1	1.2	5.4	0
Change in water volume	Ac-ft	Storage	64.7	0.2	9,387.10	-2,339.30	4	0	1,250.40	-12.9
Evaporation from water	Ac-ft	Outflow	10,669.20	29.2	68	1.7	1,168.50	3.2	16.2	0
Surface-water outflow	Ac-ft	Outflow	989	2.7	2,641.40	0.6	0	0	0	0

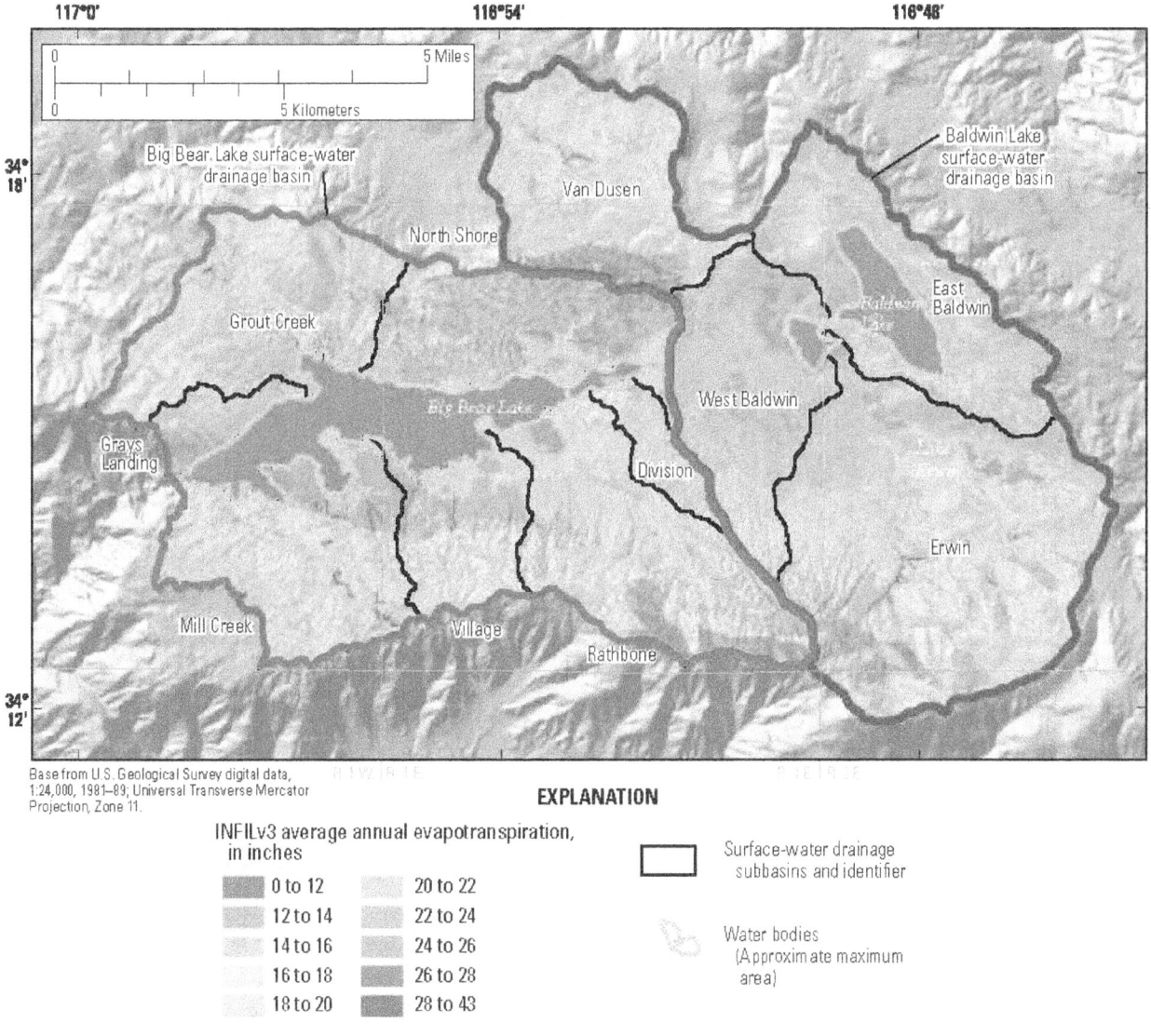

Figure 31. Simulated average annual evapotranspiration for the Big Bear study area, San Bernardino County, California, for water-years 1950–2005, from the revised climate model.

Recharge

The average annual simulated recharge in the Big Bear study area for water-years 1950–2005 was about 5,480 acre-ft/yr, with about 2,800 acre-ft/yr in the Big Bear surface-water drainage basin and about 2,680 acre-ft/yr in the Baldwin surface-water drainage basin (*table 13*). The spatial distribution of simulated 1950–2005 average annual recharge (*fig. 35*) is similar to that of lateral seepage—the pattern was primarily controlled by the permeability of the bedrock or unconsolidated deposits underlying the root zone. The greatest amount of recharge was simulated in the higher altitude regions of the North Shore and Rathbone subbasins of the Big Bear surface-water drainage basin and the Erwin subbasin in the Baldwin Lake surface-water drainage basin (*table 13; fig. 35*).

Recharge rates exceeding 50 in/yr occurred in localized areas, primarily along channels crossing high-permeability bedrock (carbonates) and unconsolidated deposits. Also, higher recharge rates tended to occur in active channels subject to high frequency or long duration of runoff (such as runoff from snowmelt) that maintained saturated conditions for sustained periods of time. Recharge rates can be relatively high (at least 5 in/yr) in locations where thin soils and sparse vegetation are underlain by high-permeability carbonates and sedimentary rock types in Erwin, Rathbone, North Shore, and Van Dusen subbasins. Recharge rates were relatively low (0.01 to 0.5 in/yr) in locations underlain by low-permeability rock types, such as granites and quartzite, within parts of Gray's Landing, Mill Creek, Grout Creek, Village, Rathbone, Erwin, Van Dusen, and East Baldwin subbasins. Locations

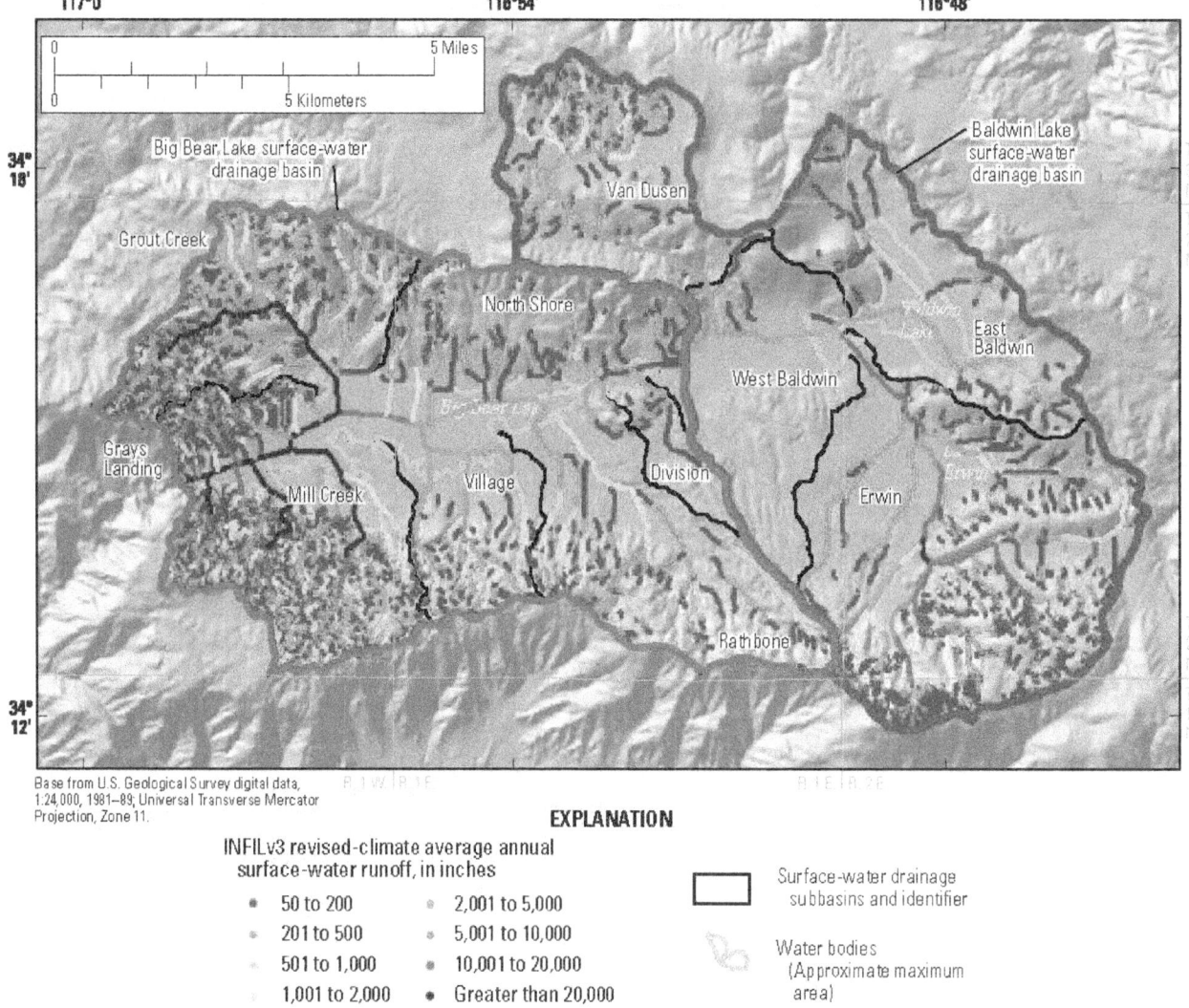

Figure 32. Simulated average annual runoff for the Big Bear study area, water-years 1950–2005, from the revised climate model, Big Bear Valley, San Bernardino County, California.

characterized by rugged topography and steep slopes were more likely to have seepage rates that are higher than recharge rates; in contrast, flat areas were more likely to have higher recharge rates and lower seepage rates.

Model Uncertainty and Limitations

The water-balance method used in the INFILv3 model incorporates several assumptions that simplify the physics of unsaturated groundwater flow. For example, the water-balance calculations assume that the process of vapor flow and the effects of temperature on water density are negligible. Water density is assumed to be constant, allowing the governing equations in the water-balance model to be applied as a volume balance rather than as a mass balance. Also, recharge is assumed to occur as gravity drainage under a unit gradient.

The effect of capillary forces on unsaturated flow in the root zone is not part of the model.

The INFILv3 model simulates streamflow originating as runoff and as subsequent overland flow and does not simulate streamflow originating as base flow from deep groundwater discharge or as through-flow from perched zones, such as the alluvium-bedrock contact in washes. INFILv3 was modified for this study to include a perched zone beneath the root zone in an effort to better simulate seepage, and ultimately recharge, in the shallow subsurface in mountainous terrain; however, groundwater discharge was not simulated. Simulation of daily streamflow in the INFILv3 model is based on a daily routing algorithm that assumes episodic streamflow with durations less than 24 hours. Simulated streamflow either discharges from the drainage basin or infiltrates into the root zone in the daily time step. Dispersive streamflow, which can be an important characteristic of streamflow and overland flow

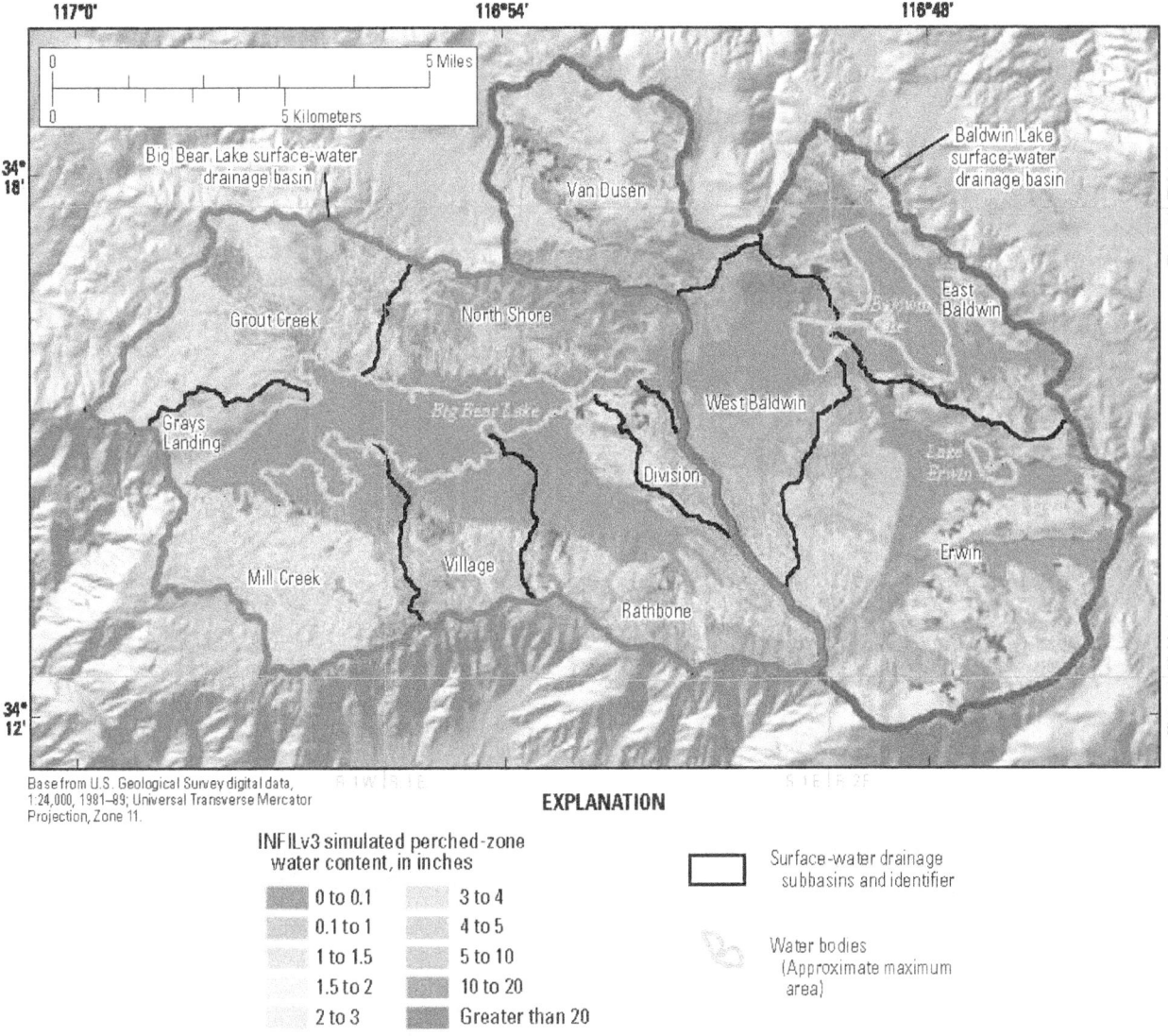

Figure 33. Simulated average annual water content in the shallow-groundwater zone for the Big Bear study area, San Bernardino County, California, water-years 1950–2005, from the revised climate model.

across alluvial fans and basins, is not directly represented in the surface-water flow-routing algorithm. All surface-water flow is simulated as convergent streamflow.

Sources of model uncertainty include input parameters, such as the hydraulic conductivity of bedrock, soil thickness, soil hydrologic properties, parameters used to define stream-channel characteristics, root-zone depth, and root density as a function of depth. For this application of INFILv3, the simulation of the shallow groundwater zone increased the number of model parameters. Additional sources of model uncertainty include model limitations representing spatial and temporal distribution of precipitation and air temperature from available climate records limitations representing runoff generation and subsequent streamflow by using assumed durations for precipitation, snowmelt, and streamflow. As shown in this study, uncertainties related to the quantity and distribution

of precipitation can make a large difference in the simulated recharge.

The INFILv3 model was calibrated to measured lake levels and estimated volumes for Big Bear and Baldwin Lakes. The use of lake levels and volumes for calibration required the use of the LAKE model to process the INFILv3 results, which in turn required many simplifying assumptions in the representation of the lake-area water-balance for each water body. Numerous LAKE model configurations, independent of the INFILv3 model configurations, were analyzed as part of the calibration process; however, the combined set of model configurations do not provide an exhaustive range of possible parameter sets for the many estimated and assumed parameters required by the models. Therefore, the model in this study with the best calibration result does not necessarily represent the best model for the study area.

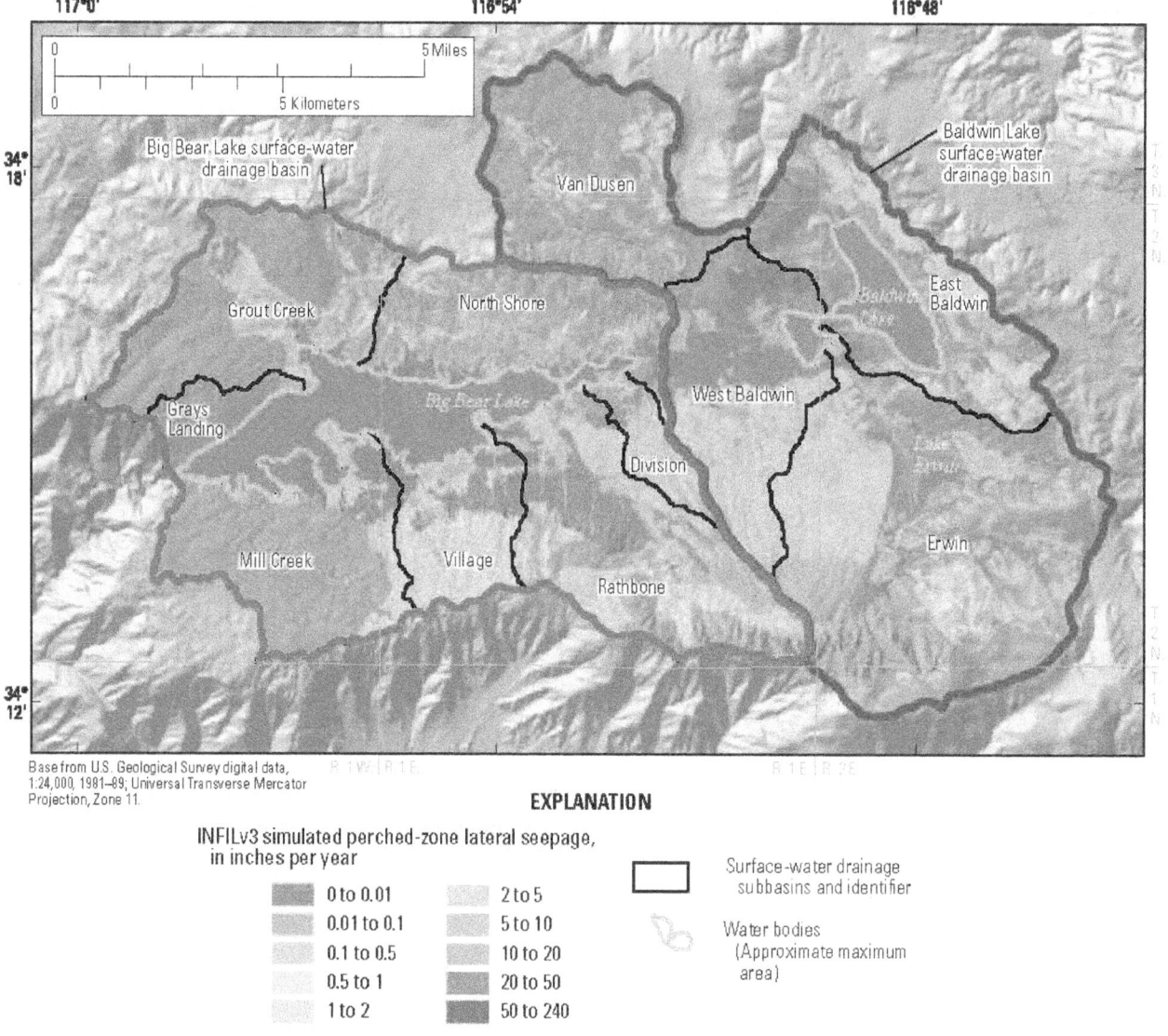

Figure 34. Simulated average annual lateral seepage in the shallow-groundwater zone for the Big Bear study area, San Bernardino County, California, for water-years 1950–2005, from the revised climate model.

Uncertainties Associated with the Shallow Groundwater Zone

The modification of INFILv3 to include a shallow groundwater zone (SGWZ) and the process of seepage decreased the simulated surface-water outflow and recharge as a result of an increase in evapotranspiration. Inclusion of the seepage component reduced recharge by as much as 50 percent for some model configurations tested in the sensitivity analysis. Although the base-case model with the simulation of a SGWZ had the best model fit, models tested in the sensitivity analysis that did not include a SGWZ adequately simulated Big Bear Lake volume. The modification to include seepage in the SGWZ primarily was done to more accurately represent conceptual models of the SGWZ developed in previous INFILv3 applications involving groundwater-flow models that

were calibrated to well data. Without data to identify locations of transient shallow perching at various locations throughout the Big Bear and Baldwin Lakes surface-water drainage basins, and without continuous hydrograph records characterizing both the overland flow and baseflow components of streamflow for at least several of the many separate drainages, a high level of uncertainty exists in the representation of the SGWZ and the process of seepage.

Uncertainties Associated with the LAKE Model

The addition of the LAKE model increased the number of estimated and assumed parameters used in the combined INFILv3-LAKE model configuration. The additional parameters adjusted during model calibration included the dry-lakebed retention and soil-zone storage capacity, the specified

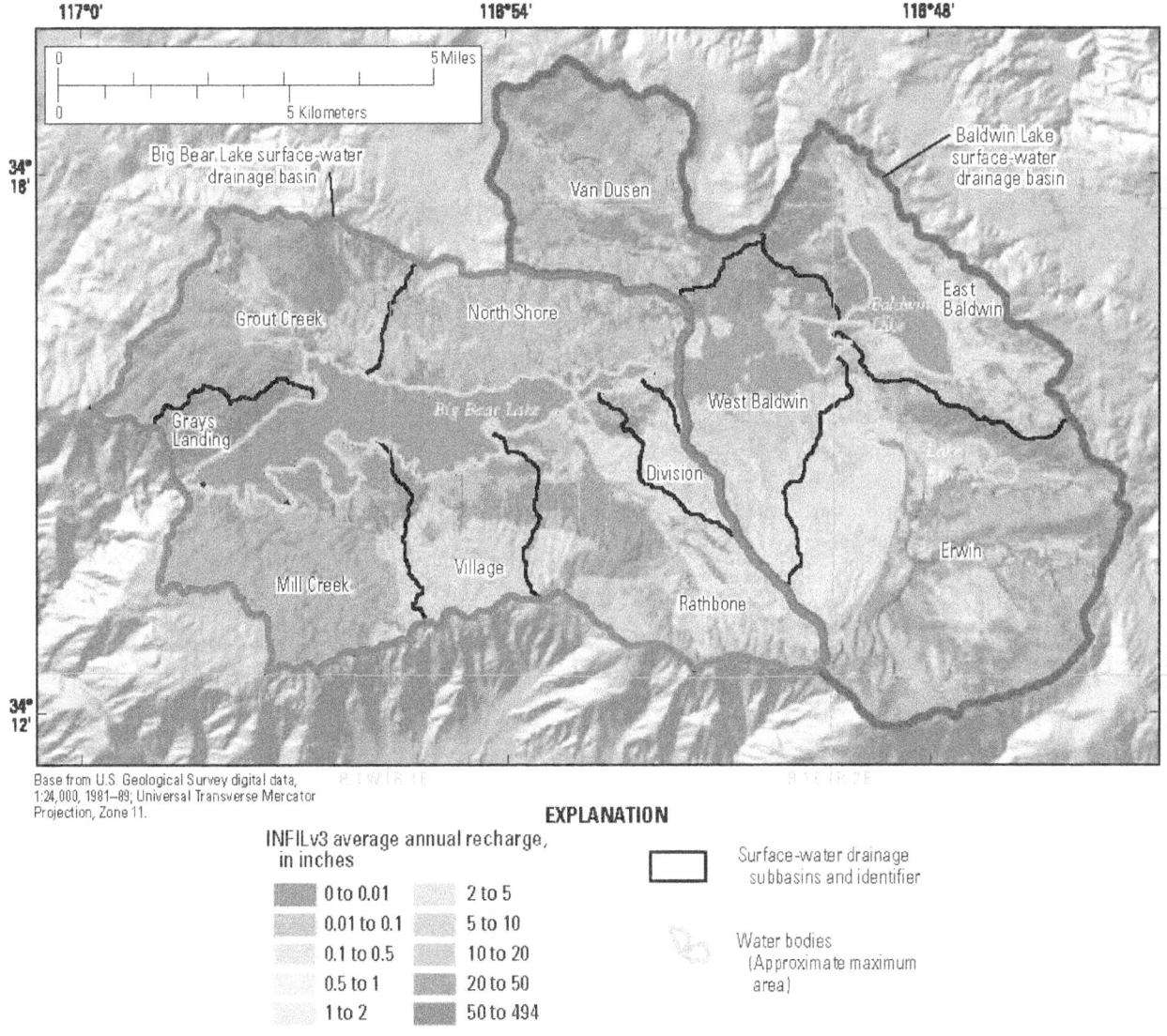

Figure 35. Simulated average annual groundwater recharge for the Big Bear study area, San Bernardino County, California, for water-years 1950–2005, from the revised climate model.

steady-state groundwater-inflow rate, the specified transient groundwater-inflow rate, the Hortonian runoff fraction, coefficients defining the relative distribution of surface-water and groundwater inflows to the wetted area and the dry lakebed, and initial conditions needed for the LAKE simulations. The LAKE model utilizes all of the INFILv3 simulated recharge as a source of groundwater discharge to the lake. Groundwater pumping would reduce the quantity of groundwater discharge to both Big Bear and Baldwin Lakes. A linked surface-water/groundwater flow model is needed to evaluate the surface-water/groundwater interactions.

Additional uncertainty in LAKE simulations for Big Bear Lake was caused by simplifying assumptions used to represent certain components of the water balance. Water losses due

to seepage through the lakebed, particularly in the proximity of Bear Valley Dam, were not included in the water balance (seepage outflow was assumed to be zero). Managed water transfers (such as the diversion of water needed for snowmaking at ski areas or diversion for irrigation) and discharge for the Bear Valley Dam were not included in the water balance because of lack of available data. The stage-discharge relation used to simulate discharge at the dam was estimated by using a very simplified representation of operations at Bear Valley Dam. Freezing and melting of the lake surface was not represented in the model.

A major source of uncertainty associated with the LAKE model developed for Baldwin Lake is the use of a DEM to estimate the level–area–volume relation. This source of uncertainty is likely to be high because Baldwin Lake is a shallow playa lake and small changes in elevation result in large relative changes in lake area and volume. Additional sources of uncertainty for the Baldwin Lake water balance include the operation of the sewage-treatment facility within the area of the playa and the development of groundwater resources (pumping).

Summary of Simulation Results

A distributed-parameter watershed model, INFILv3, was applied to the Big Bear and Baldwin Lakes surface-water drainage basins to develop spatially- and temporally-distributed estimates of recharge for the Big Bear study area. The INFILv3 model was modified to include a shallow groundwater zone (SGWZ) to simulate lateral seepage, and was calibrated by using available records of lake levels and estimated volumes for Big Bear and Baldwin Lakes. The use of lake levels and volumes for model calibration required the development of separate water-balance models for each lake area, which were used in conjunction with the INFILv3 model. The lake-area water balance model, LAKE, used output from the INFILv3 model to simulate the daily water balance for both the dry-lakebed and the wetted-lake areas.

An initial calibration to the lake volumes was attempted using a base-case watershed model. The base-case model configuration consisted of a preliminary-climate input for INFILv3 by using daily climate data from 35 NCDC stations and average monthly PRISM maps. The initial calibration provided a match of less than 1 percent difference between simulated and estimated lake volumes for Big Bear Lake, but lake volumes were overestimated for Baldwin Lake, and a simultaneous calibration to both lakes using a consistent set of model parameters could not be achieved. The calibration results for Baldwin Lake indicated that the INFILv3 base-case model simulated too much surface-water and groundwater inflow to the lake to obtain a satisfactory fit between simulated and measured lake volumes, even when assuming no groundwater inflow to the lake area. A sensitivity analysis indicated that a calibration that met the statistical goodness-of-fit criteria for Baldwin Lake was obtained only by reducing all inflows generated by the base-case INFILv3 model (precipitation, surface-water, and groundwater), indicating that the precipitation input for the base-case model was likely too high for the Baldwin Lake surface-water drainage basin.

To improve the match between measured and spatially-interpolated precipitation in the Baldwin Lake surface-water drainage basin, the climate input was revised by modifying the PRISM monthly precipitation maps. As with the climate input, the PRISM monthly precipitation maps were used to define the spatial-interpolation model for precipitation and the monthly maximum and minimum air temperature for the revised climate input. The modified PRISM monthly precipitation maps were developed by using the ratio of recorded average monthly precipitation to PRISM average monthly precipitation at 84 NCDC, RAWS, CIMIS, and SBC stations having at least 10 years of record for a given month. The adjusted PRISM data provided a better match to the drier average monthly precipitation measured in the Baldwin Lake drainage basin, resulting in an improved representation of localized rain-shadow effects. The INFILv3 revised climate model calibrated well to the Baldwin Lake record. Calibration results for Big Bear Lake, although not as good as results obtained using the base-case model, were still well within the criteria used to indicate a satisfactory calibration. The revised climate model was selected as most appropriate for estimating spatially and temporally distributed recharge and for evaluating the water balance for the Big Bear study area.

The INFILv3 simulation results obtained with the revised climate model for water-years 1950-2005 were used to develop time-averaged, spatially-distributed estimates for all components of the water balance for all subbasins in the Big Bear and Baldwin Lakes surface-water drainage basins. Total precipitation (rain and snow) for both basins was about 73,290 acre-ft/yr using the revised climate model; rain and snow were fairly evenly divided (*table 13*). Assuming a closed surface-water basin but accounting for water discharging to Big Bear and Baldwin Lakes, about 82 percent of the precipitation for the study area (excluding the areas of the lakes) was returned to the atmosphere by evapotranspiration and sublimation (about 59,540 acre-ft/yr as evapotranspiration and about 3,860 acre-ft/yr as sublimation).

A total of about 5,480 acre-ft/yr was simulated as recharge in the Big Bear study area—about 2,800 acre-ft/yr in the Big Bear Lake surface-water drainage basin and about 2,680 acre-ft/yr in the Baldwin Lake surface-water drainage basin. The simulated total recharge was about 11 percent of the total basin precipitation for the land areas upstream of the lakes. The model results indicated spatial variability coincident with variations in soil thickness and bedrock permeability; rates were high for recharge along stream channels crossing alluvium or highly permeable bedrock. For many locations, infiltration of stream flow was the dominant recharge mechanism simulated by INFILv3. The INFILv3 revised climate model simulated average recharge for water-years 1950–2005 (about 5,480) was about 35 percent of the value estimated by using the BCM for water-years 1971–2000 (about 15,800 acre-ft/yr; *tables 3* and *13*). One of the main reasons for this difference was that the precipitation simulated by the INFILv3 revised climate model (about 73,290 acre-ft/yr) was about 20 percent lower than that simulated by the BCM (about 92,050 acre-ft/yr). Another principle reason was that the INFILv3 model simulated lateral seepage, which resulted in a greater amount of evapotranspiration compared to the BCM.

Simulated surface runoff varied greatly between the Baldwin and the Big Bear Lakes surface-water drainage basins and subbasins. The calibrated revised climate model indicated that simulated inflow into Big Bear Lake was more than an order of magnitude greater than inflow into Baldwin Lake. Big Bear Lake surface-water drainage basin was dominated by runoff; in contrast, the Baldwin Lake surface-water drainage basin was dominated by recharge processes. This difference can be explained by the fact that the Baldwin Lake surface-water drainage basin has a comparatively greater percentage of thick soils and long surface-water flow paths over permeable, unconsolidated deposits, where large channel losses occur during the daily routing of streamflow, than many of the subbasins discharging into Big Bear Lake, which are characterized by thin soils and impermeable bedrock.

An additional factor contributing to increased recharge and decreased runoff in Baldwin Lake surface-water drainage basin compared to Big Bear Lake basin is the relatively high percentage of precipitation occurring as snow at the higher altitudes of the Erwin subbasin along the southern edge of the basin boundary, where simulated average snow fall was more than 57 percent of the average precipitation. Daily runoff occurring in response to snowmelt tends to be of longer duration but smaller in magnitude relative to runoff occurring in response to storms with relatively shorter durations of only several hours or days. The smaller magnitude runoff from snowmelt is more likely to infiltrate prior to reaching Baldwin Lake, thus increasing simulated recharge within the basin but decreasing surface-water inflows to the lake. In contrast, Big Bear Lake surface-water drainage basin has a higher percentage of precipitation occurring as rain at the highest locations in the basin, such as the western edge of the basin boundary. The rainfall is more likely to generate high-volume runoff over a shorter duration, and this tends to increase surface-water inflows to the lake.

Preliminary Assessment of the Source and Age of Groundwater in the Big Bear Valley

By Justin T. Kulongoski, Allen H. Christensen, and Peter Martin

In 2005 and 2006, one spring and eight wells (*fig. 36*, *table 16*) were sampled and analyzed for chemical and isotopic data to determine if isotopic techniques could be used to assess the sources and ages of groundwater in the Big Bear Valley. Samples were analyzed for the stable isotopes of oxygen (oxygen-18) and hydrogen (hydrogen-2, or deuterium)

to determine the source of water to wells and to evaluate the movement of water through the study area. Selected samples were analyzed for the radioactive isotopes of hydrogen (hydrogen-3, or tritium), carbon (carbon-14, or ^{14}C), and helium-4 (^{4}He) to determine the age, or time since recharge, of the groundwater. Borehole geophysical logs of flow velocity and temperature were collected in conjunction with depth-dependent water samples from a deep long-screened well in the Baldwin Lake area (well 2N/1E-12Q3) to help determine if the isotopic characteristics of the groundwater basin varied with depth.

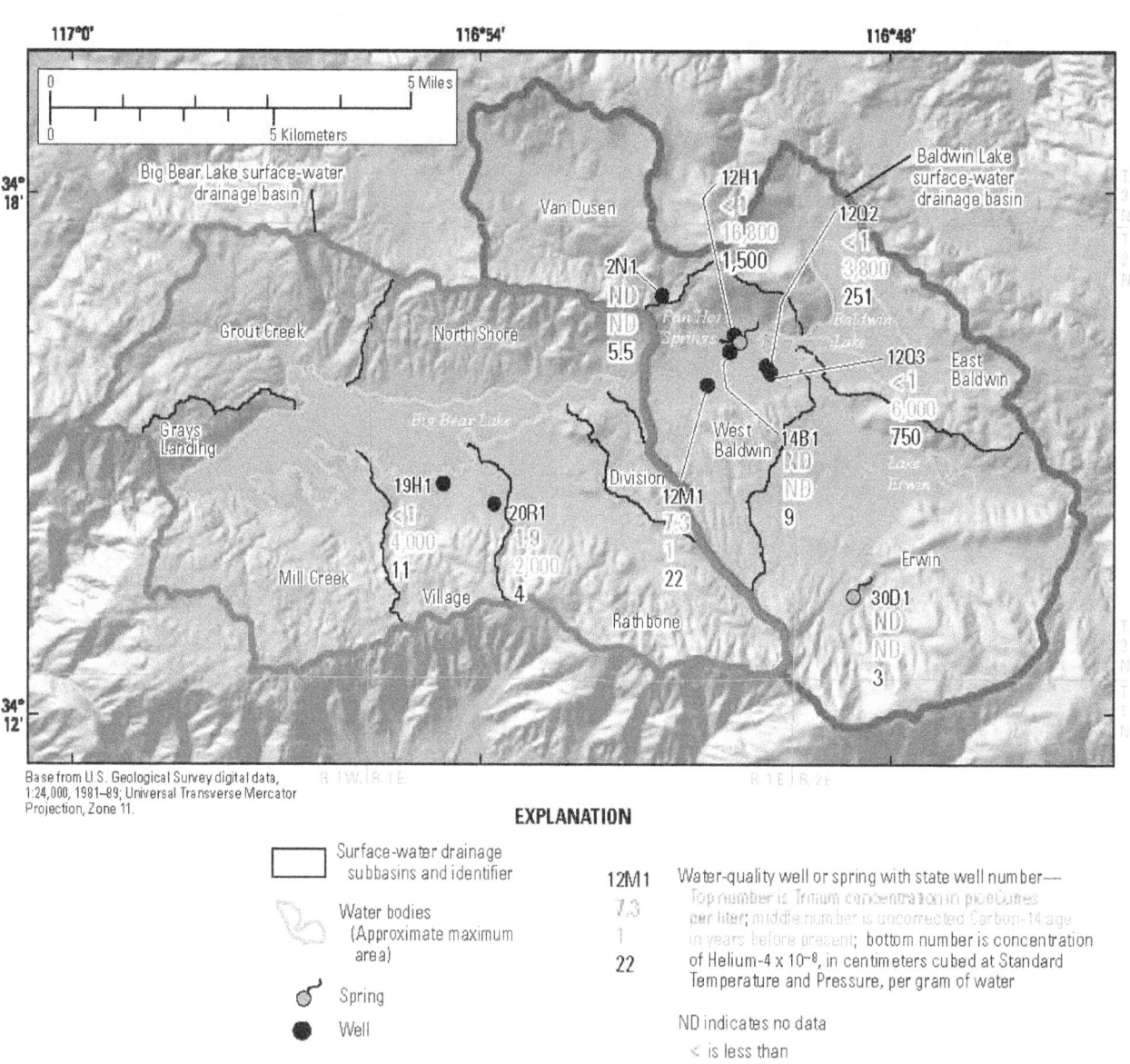

Base from U.S. Geological Survey digital data, 1:24,000, 1981–89; Universal Transverse Mercator Projection, Zone 11.

EXPLANATION

☐ Surface-water drainage subbasins and identifier

Water bodies (Approximate maximum area)

Spring

● Well

12M1
7.3
1
22   Water-quality well or spring with state well number— Top number is Tritium concentration in picocuries per liter; middle number is uncorrected Carbon-14 age in years before present; bottom number is concentration of Helium-4 x 10⁻⁸, in centimeters cubed at Standard Temperature and Pressure, per gram of water

ND indicates no data
< is less than

Figure 36.   Isotopic data for selected wells in the Big Bear Valley, San Bernardino County, California.

Table 16. Chemical and isotopic data for selected wells and springs in the Big Bear Valley, San Bernardino County, California.

[See *figure 36* for site locations. The five-digit USGS parameter code in parentheses is used to uniquely identify each water constituent or property in the NWIS database. **Abbreviations:** $CaCO_3$, calcium carbonate; ccSTP/g, cubic centimeters per gram at standard temperature and pressure; E, value estimated; mg/L, milligrams per liter; mm/dd/yyyy, month/day/year; pCi/L, picocuries per liter; per mil, per thousand; SD, spring discharge; USGS ID, U.S. Geological Survey identification number; the unique number for each site in USGS NWIS (National Water Information System) database; WD, well discharge; μg/L, micrograms per liter; μS/cm, microsiemens per centimeter; °C, degree Celsius; <, less than; –, no data]

State well number	Local identifier	USGS ID	Date (mm/dd/yyyy)	Sampling depth, (feet)	Perforated interval, (feet)	Calcium, water, filtered (mg/L) (00915)	Magnesium, water, filtered (mg/L) (00925)	Potassium, water, filtered (mg/L) (00935)	Sodium, water, filtered (mg/L) (00930)	Bromide, water, filtered (mg/L) (71870)
002N001E02N001S	Van Dusen Slant #1	341649116511701	07/28/2005	WD	–	–	–	–	–	–
002N001E12H001S	Pan Hot Springs	341618116501001	07/28/2005	WD	[1]100	4.56	E0.006	0.84	132	0.04
002N001E12H001S	Pan Hot Springs	341618116501001	11/23/2005	WD	[1]100	–	–	–	–	–
002N001E12M001S	CSD#4	341609116501801	07/26/2005	WD	40–100	57.5	32.5	2.15	11	0.07
002N001E12Q002S	CSD#3	341602116495101	07/26/2005	WD	52–316	15.5	4.77	0.89	73.4	0.04
002N001E12Q003S	CSD-3A	341600116494801	11/22/2005	WD	91–790	4.06	0.133	0.43	104	0.17
002N001E12Q003S	CSD-3A	341600116494801	03/07/2006	290	91–790	3.61	0.169	0.32	88.4	0.04
002N001E12Q003S	CSD-3A	341600116494801	03/07/2006	320	91–790	3.6	0.173	0.42	88.1	0.03
002N001E12Q003S	CSD-3A	341600116494801	03/07/2006	515	91–790	3.57	0.186	0.42	90.8	0.03
002N001E12Q003S	CSD-3A	341600116494801	03/07/2006	630	91–790	3.11	0.067	0.5	93	0.04
002N001E12Q003S	CSD-3A	341600116494801	03/07/2006	725	91–790	3.13	0.038	0.31	96.6	0.04
002N001E14B001S	CSD#9	341547116503801	07/27/2005	WD	200–516	5.54	0.027	0.41	124	0.04
002N001E19H001S	Knickerbocker	341439116543001	07/29/2005	WD	220–775	21	4.82	1.53	23.1	E0.01
002N001E20R001S	Oakwell	341423116534001	07/29/2005	WD	70–352	29.3	9.32	1.51	14.7	E0.01
002N002E30DS01S	Fish Hatchery Spring	341321116483001	07/27/2005	SD	–	–	–	–	–	–

Table 16. Chemical and isotopic data for selected wells and springs in the Big Bear Valley, San Bernardino County, California.—Continued

[See *figure 36* for site locations. The five-digit USGS parameter code in parentheses is used to uniquely identify each water constituent or property in the NWIS database. **Abbreviations:** CaCO₃, calcium carbonate; ccSTP/g, cubic centimeters per gram at standard temperature and pressure; E, value estimated; mg/L, milligrams per liter; mm/dd/yyyy, month/day/year; pCi/L, picocuries per liter; per mil, per thousand; SD, spring discharge; USGS ID, U.S. Geological Survey identification number; the unique number for each site in USGS NWIS (National Water Information System) database; WD, well discharge; µg/L, micrograms per liter; µS/cm, microsiemens per centimeter; °C, degree Celsius; <, less than; –, no data]

State well number	Local identifier	USGS ID	Date (mm/dd/yyyy)	Sampling depth, (feet)	Perforated interval, (feet)	Chloride, water, filtered (mg/L) (00940)	Fluoride, water, filtered (mg/L) (00950)	Silica, water, filtered (mg/L) (00955)	Sulfate, water, filtered (mg/L) (00945)	Residue on evaporation, dried at 180°C, water, filtered (mg/L) (70300)
							Major and selected minor ions —Continued			
002N001E02N001S	Van Dusen Slant #1	341649116511701	07/28/2005	WD	–	–	–	–	–	–
002N001E12H001S	Pan Hot Springs	341618116501001	07/28/2005	WD	[1]100	8.42	17.1	37.4	198	430
002N001E12H001S	Pan Hot Springs	341618116501001	11/23/2005	WD	[1]100	–	–	–	–	–
002N001E12M001S	CSD#4	341609116501801	07/26/2005	WD	40–100	11.3	0.14	25	29.2	337
002N001E12Q002S	CSD#3	341602116495101	07/26/2005	WD	52–316	6.54	7.46	19.3	105	282
002N001E12Q003S	CSD-3A	341600116494801	11/22/2005	WD	91–790	6.42	10.5	15.7	115	305
002N001E12Q003S	CSD-3A	341600116494801	03/07/2006	WD	91–790	5.95	9.33	15.3	90.8	276
002N001E12Q003S	CSD-3A	341600116494801	03/07/2006	290	91–790	5.95	9.38	15.1	91.2	288
002N001E12Q003S	CSD-3A	341600116494801	03/07/2006	320	91–790	6.16	9.39	15.1	99.2	285
002N001E12Q003S	CSD-3A	341600116494801	03/07/2006	515	91–790	6.16	10.1	15.6	107	286
002N001E12Q003S	CSD-3A	341600116494801	03/07/2006	630	91–790	6.46	10.1	16.1	118	307
002N001E12Q003S	CSD-3A	341600116494801	03/07/2006	725	91–790	8.64	16.7	18.8	199	409
002N001E14B001S	CSD#9	341547116503801	07/27/2005	WD	200–516	–	–	–	–	–
002N001E19H001S	Knickerbocker	341439116543001	07/29/2005	WD	220–775	2.97	0.19	18.5	12.9	153
002N001E20R001S	Oakwell	341423116534001	07/29/2005	WD	70–352	4.39	0.13	22.1	9.1	170
002N002E30DS01S	Fish Hatchery Spring	341321116483001	07/27/2005	SD	–	–	–	–	–	–

Table 16. Chemical and isotopic data for selected wells and springs in the Big Bear Valley, San Bernardino County, California.—Continued

[See *figure 36* for site locations. The five-digit USGS parameter code in parentheses is used to uniquely identify each water constituent or property in the NWIS database. **Abbreviations:** CaCO₃, calcium carbonate; ccSTP/g, cubic centimeters per gram at standard temperature and pressure; E, value estimated; mg/L, milligrams per liter; mm/dd/yyyy, month/day/year; pCi/L, picocuries per liter; per mil, per thousand; SD, spring discharge; USGS ID, U.S. Geological Survey identification number; the unique number for each site in USGS NWIS (National Water Information System) database; WD, well discharge; µg/L, micrograms per liter; µS/cm, microsiemens per centimeter; °C, degree Celsius; <, less than; —, no data]

State well number	Local identifier	USGS ID	Date (mm/dd/yyyy)	Sampling depth, (feet)	Perforated interval, (feet)	Nutrients				
						Ammonia plus organic nitrogen, water, filtered (mg/L as nitrogen) (00623)	Ammonia, water, filtered (mg/L as nitrogen) (00608)	Nitrite plus nitrate, water, filtered (mg/L as nitrogen) (00631)	Nitrite, water, filtered (mg/L as nitrogen) (00613)	Ortho-phosphate, water, filtered (mg/L as phosphorus) (00671)
002N001E02N001S	Van Dusen Slant #1	341649116511701	07/28/2005	WD	—	—	—	—	—	—
002N001E12H001S	Pan Hot Springs	341618116501001	07/28/2005	WD	[1]100	0.12	<0.04	<0.06	<0.008	<0.02
002N001E12H001S	Pan Hot Springs	341618116501001	11/23/2005	WD	[1]100	—	—	—	—	—
002N001E12M001S	CSD#4	341609116501801	07/26/2005	WD	40–100	E0.08	<0.04	1.66	<0.008	0.02
002N001E12Q002S	CSD#3	341602116495101	07/26/2005	WD	52–316	0.12	<0.04	0.11	<0.008	0.02
002N001E12Q003S	CSD-3A	341600116494801	11/22/2005	WD	91–790	E0.09	<0.04	E0.04	<0.008	<0.02
002N001E12Q003S	CSD-3A	341600116494801	03/07/2006	290	91–790	—	—	—	—	—
002N001E12Q003S	CSD-3A	341600116494801	03/07/2006	320	91–790	—	—	—	—	—
002N001E12Q003S	CSD-3A	341600116494801	03/07/2006	515	91–790	—	—	—	—	—
002N001E12Q003S	CSD-3A	341600116494801	03/07/2006	630	91–790	—	—	—	—	—
002N001E12Q003S	CSD-3A	341600116494801	03/07/2006	725	91–790	—	—	—	—	—
002N001E14B001S	CSD#9	341547116503801	07/27/2005	WD	200–516	—	—	—	—	—
002N001E19H001S	Knickerbocker	341439116543001	07/29/2005	WD	220–775	E0.06	<0.04	E0.04	<0.008	<0.02
002N001E20R001S	Oakwell	341423116534001	07/29/2005	WD	70–352	0.11	<0.04	0.12	<0.008	0.03
002N002E30DS01S	Fish Hatchery Spring	341321116483001	07/27/2005	SD	—	—	—	—	—	—

Table 16.   Chemical and isotopic data for selected wells and springs in the Big Bear Valley, San Bernardino County, California.—Continued

[See *figure 36* for site locations. The five-digit USGS parameter code in parentheses is used to uniquely identify each water constituent or property in the NWIS database. **Abbreviations:** CaCO$_3$, calcium carbonate; ccSTP/g, cubic centimeters per gram at standard temperature and pressure; E, value estimated; mg/L, milligrams per liter; mm/dd/yyyy, month/day/year; pCi/L, picocuries per liter, per mil, per thousand; SD, spring discharge; USGS ID, U.S. Geological Survey identification number; the unique number for each site in USGS NWIS (National Water Information System) database; WD, well discharge; μg/L, micrograms per liter; μS/cm, microsiemens per centimeter; °C, degree Celsius; <, less than; –, no data]

State well number	Local identifier	USGS ID	Date (mm/dd/yyyy)	Sampling depth, (feet)	Perforated interval, (feet)	Trace elements				
						Aluminum, water, filtered (μg/L) (01106)	Arsenic, water, filtered (μg/L) (01000)	Barium, water, filtered (μg/L) (01005)	Boron, water, filtered (μg/L) (01020)	Chromium, water, filtered (μg/L) (01030)
002N001E02N001S	Van Dusen Slant #1	341649116511701	07/28/2005	WD	–	–	–	–	–	–
002N001E12H001S	Pan Hot Springs	341618116501001	07/28/2005	WD	[1]100	8	2.6	E0.7	142	–
002N001E12H001S	Pan Hot Springs	341618116501001	11/23/2005	WD	[1]100	–	–	–	–	–
002N001E12M001S	CSD#4	341609116501801	07/26/2005	WD	40–100	<1.6	0.2	31.8	17	–
002N001E12Q002S	CSD#3	341602116495101	07/26/2005	WD	52–316	1.7	0.9	19.3	77	–
002N001E12Q003S	CSD-3A	341600116494801	11/22/2005	WD	91–790	10.9	3.4	2.2	104	–
002N001E12Q003S	CSD-3A	341600116494801	03/07/2006	290	91–790	12.6	3.5	1.3	89	0.43
002N001E12Q003S	CSD-3A	341600116494801	03/07/2006	320	91–790	14.3	3.5	1.6	93	0.52
002N001E12Q003S	CSD-3A	341600116494801	03/07/2006	515	91–790	18.2	3.5	30.3	92	0.37
002N001E12Q003S	CSD-3A	341600116494801	03/07/2006	630	91–790	21.5	3.9	37.5	97	0.27
002N001E12Q003S	CSD-3A	341600116494801	03/07/2006	725	91–790	26.5	3.4	22.7	98	0.19
002N001E14B001S	CSD#9	341547116503801	07/27/2005	WD	200–516	29.8	5.5	31.9	128	0.12
002N001E19H001S	Knickerbocker	341439116543001	07/29/2005	WD	220–775	1.7	19.3	72.2	32	–
002N001E20R001S	Oakwell	341423116534001	07/29/2005	WD	70–352	E1.2	1.2	54.3	E7.0	–
002N002E30DS01S	Fish Hatchery Spring	341321116483001	07/27/2005	SD	–	–	–	–	–	–

Table 16. Chemical and isotopic data for selected wells and springs in the Big Bear Valley, San Bernardino County, California.—Continued

[See *figure 36* for site locations. The five-digit USGS parameter code in parentheses is used to uniquely identify each water constituent or property in the NWIS database. **Abbreviations:** CaCO₃, calcium carbonate; ccSTP/g, cubic centimeters per gram at standard temperature and pressure; E, value estimated; mg/L, milligrams per liter; mm/dd/yyyy, month/day/year; pCi/L, picocuries per liter, per mil, per thousand; SD, spring discharge; USGS ID, U.S. Geological Survey identification number; the unique number for each site in USGS NWIS (National Water Information System) database; WD, well discharge; µg/L, micrograms per liter; µS/cm, microsiemens per centimeter; °C, degree Celsius; <, less than; –, no data]

| | | | | | | | Trace elements —Continued | | | |
State well number	Local identifier	USGS ID	Date (mm/dd/yyyy)	Sampling depth, (feet)	Perforated interval, (feet)	Iodide, water, filtered, (µg/L) (71865)	Iron, water, filtered (µg/L) (01046)	Lithium, water, filtered (µg/L) (01130)	Manganese, water, filtered (µg/L) P01056)	Strontium, water, filtered (µg/L) (01080)
002N001E02N001S	Van Dusen Slant #1	34164911651701	07/28/2005	WD	–	–	–	–	–	–
002N001E12H001S	Pan Hot Springs	34161811650100	07/28/2005	WD	[1]100	0.009	E4	84	1.3	41.1
002N001E12H001S	Pan Hot Springs	34161811650100	11/23/2005	WD	[1]100	–	–	–	–	–
002N001E12M001S	CSD#4	34160911650180	07/26/2005	WD	40–100	E0.001	E4	E1	<0.6	117
002N001E12Q002S	CSD#3	34160211649510	07/26/2005	WD	52–316	0.004	24	4	2.7	65.7
002N001E12Q003S	CSD-3A	34160011649480	11/22/2005	WD	91–790	0.007	29	7	1.5	26.2
002N001E12Q003S	CSD-3A	34160011649480	03/07/2006	WD	91–790	0.005	7	7	E0.4	22.4
002N001E12Q003S	CSD-3A	34160011649480	03/07/2006	290	91–790	0.005	<6	7	<0.6	22.7
002N001E12Q003S	CSD-3A	34160011649480	03/07/2006	320	91–790	–	14	7	E0.6	22.1
002N001E12Q003S	CSD-3A	34160011649480	03/07/2006	515	91–790	–	15	7	E0.6	19.9
002N001E12Q003S	CSD-3A	34160011649480	03/07/2006	630	91–790	0.007	8	8	0.7	19.9
002N001E12Q003S	CSD-3A	34160011649480	03/07/2006	725	91–790	–	16	13	0.6	37.6
002N001E14B001S	CSD#9	34154711650380	07/27/2005	WD	200–516	–	–	–	–	–
002N001E19H001S	Knickerbocker	34143911654300	07/29/2005	WD	220–775	E0.001	<6	2	<0.6	487
002N001E20R001S	Oakwell	34142311653400	07/29/2005	WD	70–352	<0.002	E6	3	E0.6	890
002N002E30DS01S	Fish Hatchery Spring	34132111648300	07/27/2005	SD	–	–	–	–	–	–

Table 16. Chemical and isotopic data for selected wells and springs in the Big Bear Valley, San Bernardino County, California.—Continued

[See *figure 36* for site locations. The five-digit USGS parameter code in parentheses is used to uniquely identify each water constituent or property in the NWIS database. **Abbreviations:** CaCO₃, calcium carbonate; ccSTP/g, cubic centimeters per gram at standard temperature and pressure; E, value estimated; mg/L, milligrams per liter; mm/dd/yyyy, month/day/year; pCi/L, picocuries per liter; per mil, per thousand; SD, spring discharge; USGS ID, U.S. Geological Survey identification number; the unique number for each site in USGS NWIS (National Water Information System) database; WD, well discharge; µg/L, micrograms per liter; µS/cm, microsiemens per centimeter; °C, degree Celsius; <, less than; –, no data]

State well number	Local identifier	USGS ID	Date (mm/dd/yyyy)	Sampling depth, (feet)	Perforated interval, (feet)	Field constituents			
						Dissolved oxygen, water, unfiltered (mg/L) {00300}	pH, water, unfiltered, field (standard units) {00400}	Specific conductance, water (µS/cm at 25°C) {00095}	Temperature, water, (°C) {00010}
002N001E02N001S	Van Dusen Slant #1	341649116511701	07/28/2005	WD	–	3.6	7.5	405	11
002N001E12H001S	Pan Hot Springs	341618116501001	07/28/2005	WD	¹100	2.2	9.6	676	32.5
002N001E12H001S	Pan Hot Springs	341618116501001	11/23/2005	WD	¹100	–	9.6	672	32.2
002N001E12M001S	CSD#4	341609116501801	07/26/2005	WD	40–100	8.1	7.3	559	12.5
002N001E12Q002S	CSD#3	341602116495101	07/26/2005	WD	52–316	0.2	7.5	463	21.3
002N001E12Q003S	CSD-3A	341600116494801	11/22/2005	WD	91–790	0	9.2	488	24.9
002N001E12Q003S	CSD-3A	341600116494801	03/07/2006	WD	91–790	–	9.1	449	25.9
002N001E12Q003S	CSD-3A	341600116494801	03/07/2006	290	91–790	–	9	448	25.9
002N001E12Q003S	CSD-3A	341600116494801	03/07/2006	320	91–790	–	9.1	461	25.4
002N001E12Q003S	CSD-3A	341600116494801	03/07/2006	515	91–790	–	9.2	471	25.9
002N001E12Q003S	CSD-3A	341600116494801	03/07/2006	630	91–790	–	9.3	492	25.9
002N001E12Q003S	CSD-3A	341600116494801	03/07/2006	725	91–790	–	9.5	654	25.9
002N001E14B001S	CSD#9	341547116503801	07/27/2005	WD	200–516	2	7.8	337	13
002N001E19H001S	Knickerbocker	341439116543001	07/29/2005	WD	220–775	3	8.2	237	17.5
002N001E20R001S	Oakwell	341423116534001	07/29/2005	WD	70–352	5	7.8	280	11
002N002E30DS01S	Fish Hatchery Spring	341321116483001	07/27/2005	SD	–	7.1	7.6	393	8.5

Table 16. Chemical and isotopic data for selected wells and springs in the Big Bear Valley, San Bernardino County, California.—Continued

[See *figure 36* for site locations. The five-digit USGS parameter code in parentheses is used to uniquely identify each water constituent or property in the NWIS database. **Abbreviations:** CaCO₃, calcium carbonate; ccSTP/g, cubic centimeters per gram at standard temperature and pressure; E, value estimated; mg/L, milligrams per liter; mm/dd/yyyy, month/day/year; pCi/L, picocuries per liter; per mil, per thousand; SD, spring discharge; USGS ID, U.S. Geological Survey identification number: the unique number for each site in USGS NWIS (National Water Information System) database; WD, well discharge; µg/L, micrograms per liter; µS/cm, microsiemens per centimeter; °C, degree Celsius; <, less than; –, no data]

State well number	Local identifier	USGS ID	Date (mm/dd/yyyy)	Sampling depth, (feet)	Perforated interval, (feet)	Field constituents —Continued			
						Alkalinity, water, filtered, fixed endpoint (pH 4.5) titration, field (mg/L as CaCO₃) (39036)	Alkalinity, water, filtered, incremental titration, field (mg/L as CaCO₃) (39086)	Bicarbonate, water, filtered, incremental titration, field (mg/L) (00453)	Carbonate, water, filtered, incremental titration, field (mg/L) (00452)
002N001E02N001S	Van Dusen Slant #1	341649116511701	07/28/2005	WD	–	220	217	264	0
002N001E12H001S	Pan Hot Springs	341618116501001	07/28/2005	WD	¹100	48	47	31	10
002N001E12H001S	Pan Hot Springs	341618116501001	11/23/2005	WD	¹100	–	–	–	–
002N001E12M001S	CSD#4	341609116501801	07/26/2005	WD	40–100	250	249	303	0
002N001E12Q002S	CSD#3	341602116495101	07/26/2005	WD	52–316	88	88	106	0
002N001E12Q003S	CSD-3A	341600116494801	11/22/2005	WD	91–790	74	73	76	6
002N001E12Q003S	CSD-3A	341600116494801	03/07/2006	WD	91–790	88	86	93	6
002N001E12Q003S	CSD-3A	341600116494801	03/07/2006	290	91–790	86	83	E98	E2
002N001E12Q003S	CSD-3A	341600116494801	03/07/2006	320	91–790	87	85	E93	E5
002N001E12Q003S	CSD-3A	341600116494801	03/07/2006	515	91–790	77	77	80	6
002N001E12Q003S	CSD-3A	341600116494801	03/07/2006	630	91–790	72	68	68	7
002N001E12Q003S	CSD-3A	341600116494801	03/07/2006	725	91–790	36	32	E28	E4
002N001E14B001S	CSD#9	341547116503801	07/27/2005	WD	200–516	160	163	197	0
002N001E19H001S	Knickerbocker	341439116543001	07/29/2005	WD	220–775	110	112	134	1
002N001E20R001S	Oakwell	341423116534001	07/29/2005	WD	70–352	140	137	166	0
002N002E30DS01S	Fish Hatchery Spring	341321116483001	07/27/2005	SD	–	200	200	243	0

Table 16. Chemical and isotopic data for selected wells and springs in the Big Bear Valley, San Bernardino County, California.—Continued

[See *figure 36* for site locations. The five-digit USGS parameter code in parentheses is used to uniquely identify each water constituent or property in the NWIS database. **Abbreviations:** $CaCO_3$, calcium carbonate; ccSTP/g, cubic centimeters per gram at standard temperature and pressure; E, value estimated; mg/L, milligrams per liter; mm/dd/yyyy, month/day/year; pCi/L, picocuries per liter; per mil, per thousand; SD, spring discharge; USGS ID, U.S. Geological Survey identification number: the unique number for each site in USGS NWIS (National Water Information System) database; WD, well discharge; μg/L, micrograms per liter; μS/cm, microsiemens per centimeter; °C, degree Celsius; <, less than; –, no data]

State well number	Local identifier	USGS ID	Date (mm/dd/yyyy)	Sampling depth, (feet)	Perforated interval, (feet)	Deuterium/ protium ratio, water, unfiltered, (per mil) (82082)	Oxygen-18/ oxygen-16 ratio, water, unfiltered, (per mil) (82085)	Tritium, water, unfiltered (pCi/L) (070000)	Tritium 2-sigma combined uncertainty, water, unfiltered, (pCi/L) (75985)
002N001E02N001S	Van Dusen Slant #1	341649116511701	07/28/2005	WD	–	-84.6	-11.9	–	–
002N001E12H001S	Pan Hot Springs	341618116501001	07/28/2005	WD	¹100	-99.7	-13.66	-0.32	0.58
002N001E12H001S	Pan Hot Springs	341618116501001	11/23/2005	WD	¹100	–	–	–	–
002N001E12M001S	CSD#4	341609116501801	07/26/2005	WD	40–100	-85.2	-11.78	7.33	0.58
002N001E12Q002S	CSD#3	341602116495101	07/26/2005	WD	52–316	-94.4	-12.93	0.48	0.58
002N001E12Q003S	CSD-3A	341600116494801	11/22/2005	WD	91–790	-95.3	-12.97	0.26	0.58
002N001E12Q003S	CSD-3A	341600116494801	03/07/2006	WD	91–790	-91.7	-12.88	–	–
002N001E12Q003S	CSD-3A	341600116494801	03/07/2006	290	91–790	-92.3	-12.87	–	–
002N001E12Q003S	CSD-3A	341600116494801	03/07/2006	320	91–790	-92.7	-12.91	–	–
002N001E12Q003S	CSD-3A	341600116494801	03/07/2006	515	91–790	-92.6	-12.96	–	–
002N001E12Q003S	CSD-3A	341600116494801	03/07/2006	630	91–790	-93	-13	–	–
002N001E12Q003S	CSD-3A	341600116494801	03/07/2006	725	91–790	-98.3	-13.57	–	–
002N001E14B001S	CSD#9	341547116503801	07/27/2005	WD	200–516	-87.1	-12.14	–	–
002N001E19H001S	Knickerbocker	341439116543001	07/29/2005	WD	220–775	-81.5	-11.61	0.19	0.58
002N001E20R001S	Oakwell	341423116534001	07/29/2005	WD	70–352	-80.6	-11.57	1.95	0.58
002N002E30DS01S	Fish Hatchery Spring	341321116483001	07/27/2005	SD	–	-87.9	-12.42	–	–

Table 16. Chemical and isotopic data for selected wells and springs in the Big Bear Valley, San Bernardino County, California.—Continued

[See *figure 36* for site locations. The five-digit USGS parameter code in parentheses is used to uniquely identify each water constituent or property in the NWIS database. **Abbreviations:** CaCO₃, calcium carbonate; ccSTP/g, cubic centimeters per gram at standard temperature and pressure; E, value estimated; mg/L, milligrams per liter; mm/dd/yyyy, month/day/year; pCi/L, picocuries per liter; per mil, per thousand; SD, spring discharge; USGS ID, U.S. Geological Survey identification number: the unique number for each site in USGS NWIS (National Water Information System) database; WD, well discharge; μg/L, micrograms per liter; μS/cm, microsiemens per centimeter; °C, degree Celsius; <, less than; –, no data]

State well number	Local identifier	USGS ID	Date (mm/dd/yyyy)	Sampling depth, (feet)	Perforated interval, (feet)	Stable and radioactive isotopes —Continued			
						Carbon-14, water, filtered (percent modern) (49933)	Uncorrected C-14 age (years)	Carbon-13/ Carbon-12 ratio, water, unfiltered, (per mil) (82081)	Helium-4, water unfiltered, (ccSTP/g)
002N001E02N001S	Van Dusen Slant #1	34164911651170 1	07/28/2005	WD	–	–	–	–	5.55E–08
002N001E12H001S	Pan Hot Springs	34161811650100 1	07/28/2005	WD	¹ 100	12.46	17,200	–10.51	leaked
002N001E12H001S	Pan Hot Springs	34161811650100 1	11/23/2005	WD	¹ 100	–	–	–	1.51E–05
002N001E12M001S	CSD#4	34160911650180 1	07/26/2005	WD	40–100	96.93	modern	–13.29	2.16E–07
002N001E12Q002S	CSD#3	34160211649510 1	07/26/2005	WD	52–316	60.36	4,150	–12.05	2.51E–06
002N001E12Q003S	CSD-3A	34160011649480 1	11/22/2005	WD	91–790	45.56	6,500	–11.6	7.50E–06
002N001E12Q003S	CSD-3A	34160011649480 1	03/07/2006	WD	91–790	–	–	–	–
002N001E12Q003S	CSD-3A	34160011649480 1	03/07/2006	290	91–790	–	–	–	–
002N001E12Q003S	CSD-3A	34160011649480 1	03/07/2006	320	91–790	–	–	–	–
002N001E12Q003S	CSD-3A	34160011649480 1	03/07/2006	515	91–790	–	–	–	–
002N001E12Q003S	CSD-3A	34160011649480 1	03/07/2006	630	91–790	–	–	–	–
002N001E12Q003S	CSD-3A	34160011649480 1	03/07/2006	725	91–790	–	–	–	–
002N001E14B001S	CSD#9	34154711650380 1	07/27/2005	WD	200–516	–	–	–	9.35E–08
002N001E19H001S	Knickerbocker	34143911654300 1	07/29/2005	WD	220–775	56.28	4,750	–13.59	1.13E–07
002N001E20R001S	Oakwell	34142311653400 1	07/29/2005	WD	70–352	75.96	2,250	–13.98	4.10E–08
002N002E30DS01S	Fish Hatchery Spring	34132111648300 1	07/27/2005	SD	–	–	–	–	3.46E–08

¹ Reported depth of well.

Sample Collection and Analysis

Sample Collection of Well and Spring Discharge

Water-quality samples were collected from one spring and eight wells for this study. Sampling procedures followed protocols described in the USGS field manual (U.S. Geological Survey, 2006), unless indicated otherwise, in which case additional details are provided. Air and water temperature were determined in the field at the time sample collection began. Specific conductance, pH, and alkalinity were measured in the field. Unfiltered-water samples for analysis of tritium (^{3}H) were collected in one-liter polyethylene bottles. Unfiltered samples for analysis of stable isotopes in water were collected in small glass vials. The stable isotope and tritium bottles were sealed with a conical plastic screw cap and taped with electrical tape to preclude leakage and evaporation prior to analysis. Water for analysis of helium-4 was collected from the well or spring discharge using clear tubing connected to prepared annealed copper tubes. Following flushing, the copper tube was crimped with a cold-welder, capturing a 14 cm^3 aliquot of groundwater in the sealed tube. The clear tubing allowed the water samples to be visually inspected for air bubbles prior to crimping. Collecting fluid samples in copper tubing allows sample storage for long periods without risk of compromising gas integrity.

Depth-Dependent Water-Quality Sampling at Well 2N/1E-12Q3

In March 2006, borehole flow-velocity data were collected in conjunction with depth-dependent water samples from a deep long-screened well near Baldwin Lake (well 2N/1E-12Q3) to help determine if the isotopic characteristics of the groundwater basin varied with depth (*fig. 37*). Borehole temperature data also were collected from the well to help select water-quality sampling depths. The logs were collected from land surface to the bottom of the well, approximately 810 ft below land surface (bls). The well was flowing under artesian pressure at approximately 225 gallons per minute (gpm) when the logs were collected.

Velocity data (*fig. 37*) were collected by using an electromagnetic (EM) flowmeter. The EM flowmeter measures velocities according to Faraday's Law: the voltage generated by the movement of charged ions in water flowing through an induced magnetic field is proportional to the velocity of water flowing through the field. The tool has a flow-detection range of 0.3 to 260 feet per minute and is suitable for small velocities in unpumped wells and large velocities in pumped wells (Newhouse and others, 2005). The borehole flow was calculated by multiplying the measured velocity by the cross-sectional area of the borehole. The cross-sectional area of the

borehole was assumed to be constant; however, changes in the cross-sectional area caused by well construction (differences in the casing diameter between the blank and perforated sections) or well encrustation result in small variations in the calculated borehole flow that are not the result of flow entering or leaving the borehole. The calculated borehole flow indicates that when the well was allowed to flow under artesian conditions, the lower perforated interval (630–790 ft bls) contributed 30 percent (67.5 gpm), the middle perforated interval (295–530 ft bls) contributed 60 percent (135 gpm), and the upper perforated interval (130–250 ft bls) contributed 10 percent (22.5 gpm) of the total discharge from the well (*fig. 37*).

Analysis of the temperature log indicated that the water temperature ranged from 30.6°C (87.1°F) at the bottom of the well to 26.3°C (79.3°F) in the surface discharge (*fig. 37*). The temperature in the lower perforated interval was relatively constant at the bottom of the well (790–725 ft bls); however, from 725 to 630 ft bls, the temperature decreased from about 30.2 to 28.1°C (86.4 to 82.6°F) with a temperature gradient of 0.02°C per ft (0.040°F per ft). The change in temperature indicates that cooler water was entering the well in the lower perforated interval above 725 ft bls, as was indicated on the borehole-flow log (*fig. 37*). The temperature for the middle perforated interval ranged from 27.4°C (81.3°F) at the bottom of the interval to 26.4°C (79.5°F) at the top, with a temperature gradient of 0.004°C per ft (0.007°F per ft). The upper perforated interval had a constant temperature of about 26.3°C (79.3°F).

Water-quality samples were collected at the well discharge and at five depths in the borehole (290, 320, 515, and 725 ft bls) while the well was flowing. Depths were selected on the basis of the flow-velocity log, lithologic log, temperature log, geophysical log, and well-construction data. The samples from the borehole were collected by using a commercially available, small-diameter gas-displacement pump and procedures described by Izbicki and others (2004). Samples were analyzed at the USGS Central Laboratory for major ions and stable isotopes of hydrogen and oxygen; however, only stable isotope results are discussed in this report.

Because the well was artesian, the samples collected from the well discharge and at specific depths from within the borehole represent a composite of water that entered the well from perforated intervals below the sample-collection depth. The quality of water entering the well between selected depths (zonal flow) was estimated by coupling velocity log and depth-dependent water-quality data (Izbicki and others, 2004). By measuring the concentrations of a constituent at two sequential depths in the well, the concentration of the water entering the well in the zone between the two sample depths (Ca) was calculated as follows:

$$C_a = [(C_1Q_1 - C_2Q_2)/(Q_1 - Q_2)]$$ (3)

where

$C_1 \text{ and } C_2$ are depth-dependent
    concentrations at the lower and upper
    borehole sample locations respect-
    ively, in milligrams per liter; and

$Q_1 \text{ and } Q_2$ are borehole flow rates at the lower
    and upper locations of the sampled
    interval within the borehold estimated
    from the velocity log and the diameter
    of the well, in gallons per minute.

Laboratory Analysis

Several laboratories performed chemical and isotopic analyses on water samples collected for this study. Major ions, minor and trace elements, and nutrients were analyzed by the USGS National Water-Quality Laboratory (NWQL; Timme, 1995) by various methods as described in Faires (1993), Fishman and Friedman (1989), Fishman (1993), McLain (1993), Garbarino (1999), Garbarino and others (2006), Patton and Krysalla (2003), Patton and Truitt (1992 and 2000), and Struzeski and others (1996). Tritium activity was measured by the NRP Tritium-Light Isotope laboratory in Menlo Park, California, by electrolytic enrichment using glass cells with electodes of nickel and stainless steel (Ostlund and Werner, 1962). The electrolyzed samples were then counted in liquid scintillation counters as 1:1 mixtures of water and a commercial scintillator

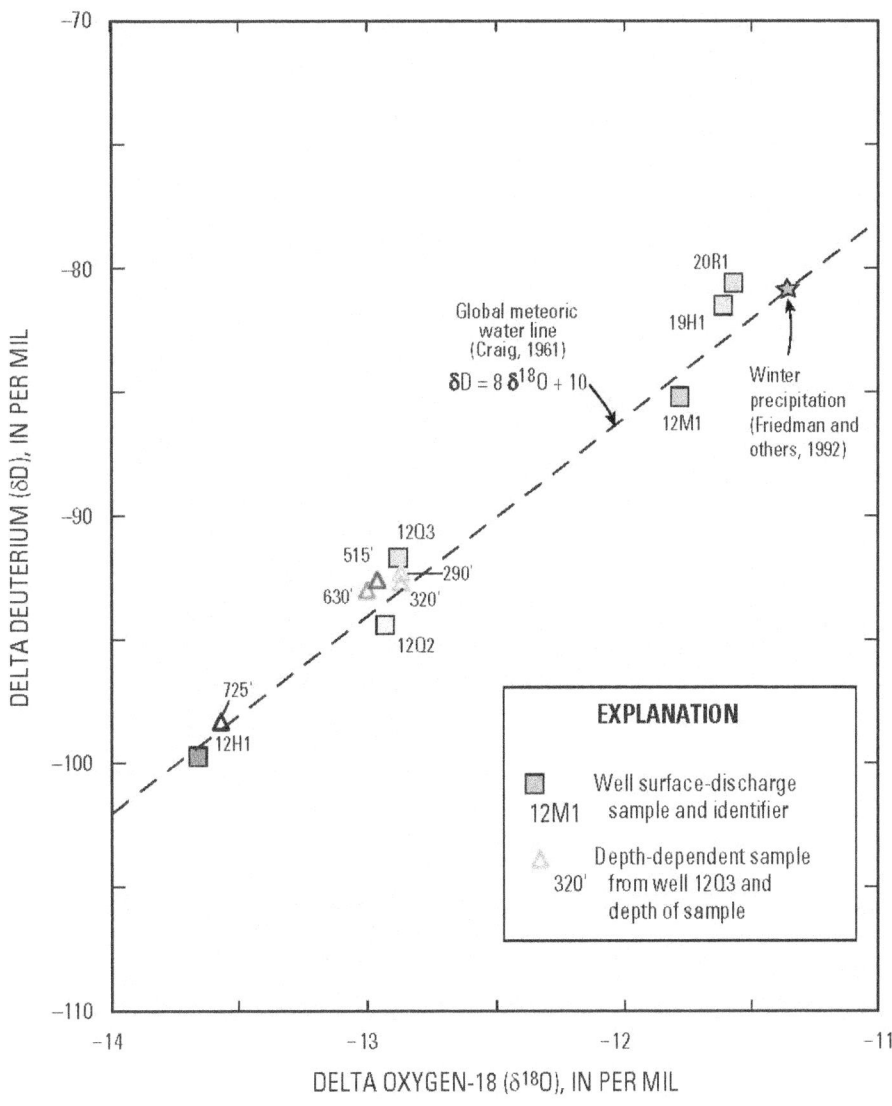

Figure 37. Velocity, temperature, and depth-dependent delta-deuterium data for well 2N/1E-12Q3 near Baldwin Lake, Big Bear Valley, San Bernardino County, California.

(Thatcher and others, 1977). Stable hydrogen and oxygen isotopes of water were analyzed by the National Research Program Stable-Isotope laboratory in Reston, Virginia (Epstein and Mayeda, 1953; Coplen and others, 1991; Coplen, 1994). Stable carbon isotopes and carbon-14 were analyzed at the University of Waterloo (Donahue and others, 1990; Jull and others, 2004). Samples collected for helium-4 were analyzed at the Fluids and Volatiles Laboratory of the Scripps Institution of Oceanography (Kulongoski and Hilton, 2002).

Stable Isotopes of Oxygen and Hydrogen

Oxygen-18 (^{18}O) and deuterium (D) are naturally occurring stable isotopes of oxygen and hydrogen. The isotopic ratios are expressed in delta notation (δ) as parts per thousand (per mil) differences relative to the standard known as Vienna Standard Mean Ocean Water (VSMOW) (Gonfiantini, 1978). The δ^{18}O and δD composition of precipitation throughout the world is linearly correlated because most of the world's precipitation is derived originally from the evaporation of seawater. This linear relationship is known as the global meteoric water line (Craig, 1961). Differences in isotopic composition can be used to help determine general atmospheric conditions at the time of precipitation and the effects of evaporation before water entered the groundwater system (Dansgaard, 1964). The δ^{18}O and δD values for groundwater relative to the global meteoric water line (MWL) provide evidence of the source of the water and fractionation processes that affected stable-isotope values. For example, water from an air mass that condensed at higher altitudes, cooler temperatures, or both, contains a greater amount of the lighter isotopes of oxygen and hydrogen and, therefore, has lighter δ^{18}O and δD values (more negative) than water that condensed from the same air mass at lower altitudes and (or) warmer temperatures. The temperature effect, which is based on measurements from North America and Europe, is -0.7 per mil/°C for δ^{18}O and -5.6 per mil/°C for δD (Dansgaard, 1964). The altitude effect, which is based on measurements taken on the western flank of the Sierra Mountain Range, is -2.3 per mil/km for δ^{18}O (Ingraham and Taylor, 1991, Rose and others, 1996). This is equivalent to -18.4 per mil/km (-0.56 per mil/100 ft) for δD, if isotope data fall on the global MWL or on a local line that is parallel to the MWL. In some areas, fractionation during atmospheric condensation and deposition, or during infiltration and groundwater recharge can result in waters yielding different δ^{18}O and δD values. Fractionation can occur through evaporation, preferential use by plants, or exchange with interstitial components of the sediment and rock matrix. Information about the source and evaporative history of water can be used to evaluate the movement of water between aquifers. Because groundwater moves slowly, isotopic data collected near the end of long flow paths typically preserve a record of groundwater recharge and movement under predevelopment conditions. This is especially useful in areas where traditional hydrologic data (such as water levels) have been altered by human activities.

The δ^{18}O and δD composition of groundwater samples collected from the study area ranged from -11.78 to -13.66 and -80.6 to -99.7 per mil, respectively (*fig. 38; table 16*). Most of the groundwater samples plot near the MWL, indicating that groundwater recharge was not subjected to evaporation before infiltrating (*fig. 38*). Partial evaporation of precipitation or runoff causes fractionation of δ^{18}O and δD that results in a shift in isotopic values to the right of the MWL. Fluid-rock interactions can result in a decrease in the δ^{18}O, and the fluid can plot slightly to the left of the MWL.

Seven years of 6-month seasonal (summer and winter) precipitation monitoring, for water-years 1982–89, reveal a weighted mean of -77 per mil for the entire period, -57 per mil for the summer period and -81 per mil for the winter period, from a fixed station at an altitude of 6,752 ft (2,055 m) near Big Bear City Airport (Friedman and others, 1992; Smith and others, 1992; and Gleason and others, 1994). Both δD and δ^{18}O data were obtained for only 2 of the 7 years, 1986 and 1987, and the weighted means for this short duration are -84.7 per mil and -11.42 per mil, respectively (calculated from data in Friedman and others, 1992). These data would plot very nearly on the global MWL; therefore, it is reasonable to impute values of δ^{18}O if oxygen isotopes had been analyzed for the entire 7-year period of precipitation monitoring (10.88 per mil, 11.38 per mil, and 8.38 per mil for the entire period, summer, and winter, respectively).

The isotopic range of δD in groundwater sampled from wells 2N/1E-20R1, 2N/1E-19H1, and 2N/1E-12M1 is similar to the volume-weighted average of winter precipitation (-77 per mil) collected near Big Bear, California (Friedman and others, 1992) suggesting that the source of groundwater recharge in these wells is precipitation from winter storms (fig. 38). The slightly lighter (more negative) values of δD and δ^{18}O in these wells could indicate that the recharge zone is at a higher altitude than Friedman's (1992) precipitation location on the valley floor at the Big Bear City Airport (6,752 ft).

The δD in the groundwater samples from wells 2N/1E-12Q2, 2N/1E-12Q3 and thermal well 2N/1E-12H1 are significantly lighter (-91.7 to -99.7 per mil) than the volume-weighted average of winter precipitation (-77 per mil) collected near the Big Bear City Airport, indicating that recharge for these wells occurred at higher altitudes or cooler conditions than modern winter precipitation on the valley floor (*fig. 38*). Uncorrected carbon-14 dates, presented in the "Carbon-14" section of this report, indicate recharge to well 2N/1E-12H1 dates to near the end of the last North American glaciation, when it likely was colder or wetter, or both, in the study area, which would cause isotope ratios to be lighter (more negative) than modern precipitation.

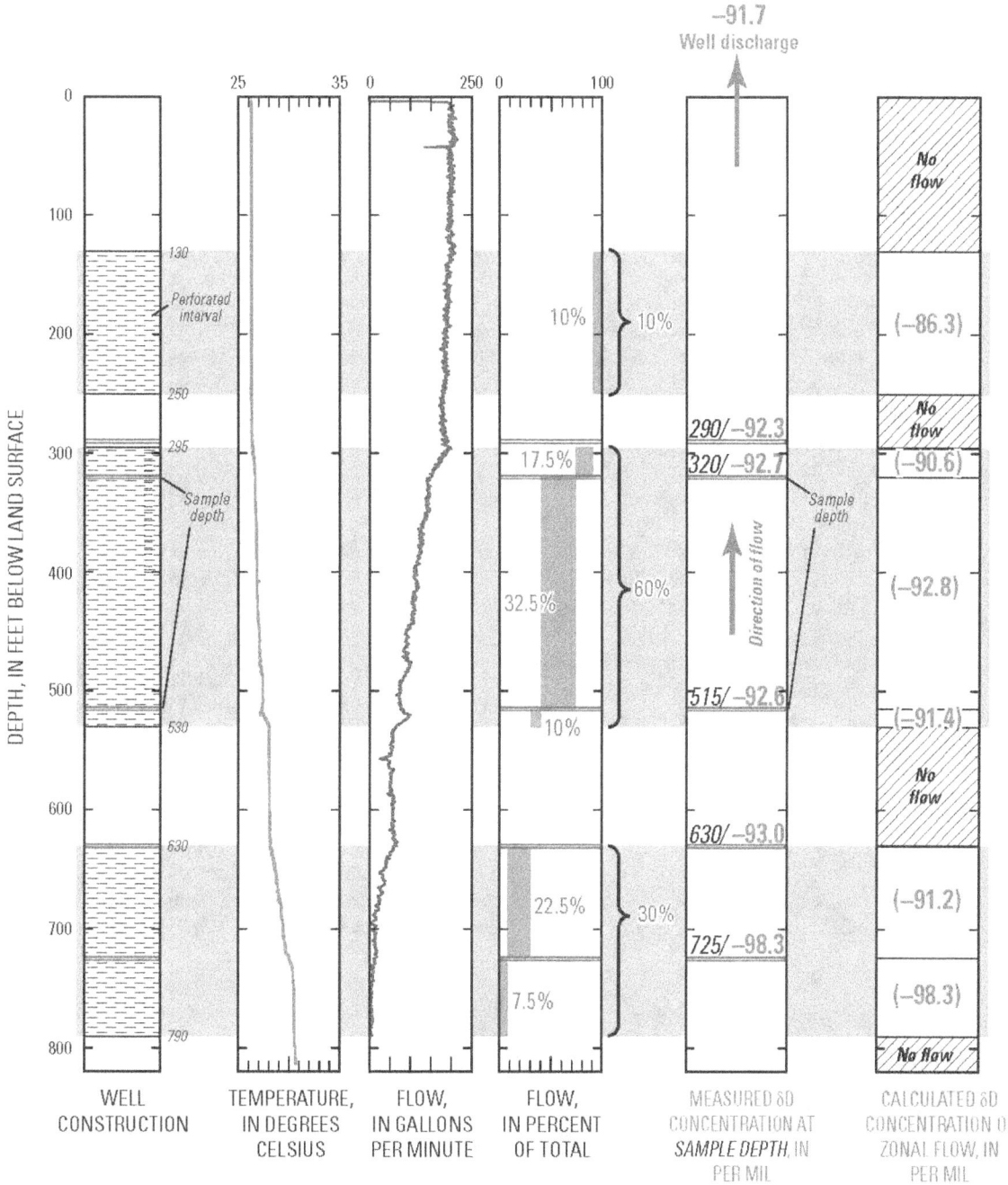

Figure 38. Relation between delta-deuterium and delta oxygen-18 from groundwater sampled in the Big Bear Valley, San Bernardino County, California.

Among the depth-dependent samples from well 2N/1E-12Q3, the δD and δ^{18}O are significantly lighter (more negative) in the sample collected at 725 ft than in the samples, collected at shallower depths in the well. The composite depth-dependent samples collected from 2N/1E-12Q3 at sample depths above 725 ft (290, 320, 525, and 630 ft) and the nearby well 2N/1E-12Q2, have similar δD and δ^{18}O, suggesting a similar source of recharge (fig. 37). The δD and δ^{18}O values from 2N/1E-12Q3 at 725 ft bls are similar to δD and δ^{18}O values from the nearby thermal well (2N/1E-12H1), indicating a similar source of recharge for both wells. The elevated temperature of water measured opposite the deep zone of well 2N/1E-12Q3 (30.6°C; *fig. 38*) and the discharge from well 2N/1E-12H1 (32.56°C; *table 16*) also suggest a similar source of water.

The source of water for the spring and wells sampled for this study is believed to be the infiltration of precipitation in the mountains that surround the Big Bear Valley. Infiltration takes place over a range of altitudes in the mountains, but water isotopes can be used to calculate the "average" altitude at which recharge occurs for samples with values falling on or near the MWL. The method extrapolates from the isotopic composition of precipitation station at the Big Bear City Airport by using the altitude effect of isotope ratios measured on the western flank of the Sierra Mountain Range (Ingraham and Taylor, 1991, Rose and others, 1996), presented earlier in this section of this report, to calculate the average altitude of recharge for a sample. The results presented in this report were obtained by substituting δD values of the well and spring samples in the following equation. An analogous calculation also could be done using δ^{18}O values of the samples, and it would yield an identical result.

$$E = E_{BBP} + \left(\frac{\delta_{s,w} - \delta_{BBP}}{Z} \right) \qquad (4)$$

where

E is the calculated elevation of recharge to a spring or well,

E_{BBP} is the elevation of the Big Bear City Airport precipitation station (6,742 ft),

$\delta_{s,w}$ is the hydrogen-isotope ratio for the spring(s) or well (w),

δ_{BBP} is the hydrogen-isotope ratio for the Big Bear City Airport precipitation station (−77 per mil), and

Z is the effect of altitude on isotopic composition of precipitation (assumed to be −0.56 per mil/100 ft, as dicussed earlier).

The altitude of recharge calculated from the δD values ranges from about 7,380 to 10,800 ft, indicating that most of the recharge occurs on the flanks of the mountains that surround the valley floor. The lowest calculated altitude of recharge was for well 2N/1E-20R1, which has perforations in the shallow alluvial deposits (less than 100 ft deep). The highest calculated altitude of recharge was for thermal well 2N/1E-12H1, near Pan Hot Springs. The maximum calculated altitude of recharge is higher than maximum altitude of the San Bernardino Mountains that form the southern boundary of the watershed (10,243 ft; *fig 2*). The calculated altitude of recharge could overestimate the actual altitude of recharge for samples from wells 2N/1E-12Q2, 2N/1E-12Q3 and 2N/1E-12H1 because water ages from these samples, described in the "Carbon-14" section of this report, indicate recharge dates to near the end of the last North American glaciation, when it likely was colder and(or) wetter in the study area, which would cause isotope ratios to be lighter (more negative) than in modern precipitation. In any case, the isotopic data indicate that most of the recharge occurs at a higher altitude than the valley floor. These results support the INFILv3 results, presented earlier in this report (*fig. 35*).

Tritium

Tritium (^{3}H) is a naturally occurring radioactive isotope of hydrogen that has a half-life of 12.4 years. The concentration of tritium is measured in picocuries per liter (pCi/L) and often reported in Tritium Units (TU), one of which equals 3.19 pCi/L. Approximately 800 kilograms of tritium were released into the atmosphere as a result of the atmospheric testing of nuclear weapons between 1952 and 1962 (Michel, 1976). As a result, tritium concentrations in precipitation and groundwater recharge increased during that time. Tritium concentrations are not affected significantly by chemical reactions other than radioactive decay because tritium is part of the water molecule. Therefore, tritium is an excellent tracer of the movement and relative age of water up to about 50 years before present (post 1952). In this report, groundwater that has detectable tritium (greater than 1.0 pCi/L) is interpreted to be water recharged after 1952, or modern recharge.

Samples collected from wells 2N/1E-20R1 and 2N/1E-12M1 had tritium concentrations in excess of 1.0 pCi/L, ranging from 1.9 to 7.3 pCi/L, indicating that these wells have received recharge within the past 50 years. Both of these wells have perforations in the shallow alluvial deposits (less than 100 ft deep), providing a pathway for local modern recharge. Tritium concentrations in samples from wells 2N/1E-19H1, 2N/1E-12Q2, and 2N/1E-12Q3, and thermal well 2N/1E-12H1 were less than or equal to 1.0 pCi/L, indicating that water from these sites was recharged before 1952 (*table 16, fig. 36*).

Carbon-14

Carbon-14 (^{14}C) is a naturally occurring radioactive isotope of carbon that has a half-life of about 5,730 years (Mook, 1980). Carbon-14 data are expressed as percent modern carbon (pmc) by comparing ^{14}C activities to the specific activity of National Bureau of Standards oxalic acid: 13.56 disintegrations per minute per gram of carbon in the year 1950 equals 100 pmc (Kalin, 2000). Carbon-14 was produced, as was tritium, by the atmospheric testing of nuclear weapons (Mook, 1980). As a result, ^{14}C activities may exceed 100 pmc in areas where groundwater contains tritium. Carbon-14 activities are used to determine the age of a groundwater sample up to more than 20,000 years before present. Carbon-14 is not part of the water molecule, and therefore, ^{14}C activities can be affected by chemical reactions that remove or add carbon to solution. In addition, ^{14}C activities are affected by mixing younger water that has high ^{14}C activity with older water that has low ^{14}C activity.

The ^{14}C activity in the six samples collected for this study ranged from about 97 to 12 pmc, which corresponds to uncorrected groundwater ages (residence times) from modern to 17,200 years old (*table 16; fig. 36*). Carbon-14 ages presented in this report do not account for changes in ^{14}C activity resulting from chemical reactions or mixing and, therefore, are considered uncorrected ages. Exchange between aqueous dissolved inorganic carbon and radiocarbon-dead carbonate in soils during movement of the groundwater will cause calculated ^{14}C ages to overstate the actual time since recharge (Vogel and Ehhalt, 1963; Brinkmann and others, 1959). The effect of this exchange with radiocarbon-dead carbon was estimated by using δ^{13}C values in the samples. The δ^{13}C was about −13.3 per mil in well 2N/1E-12M1, where the presence of tritium (^{3}H) indicates a very young age, and about −10.5 per mil in well 2N/1E-12H1 near Pan Hot Springs, where the uncorrected carbon-14 age is 17,200 years. Assuming the values for δ^{13}C represent the isotope ratio soon after the time of recharge to groundwater sampled at 2N/1E-12H1, and that the isotope ratio from this sample has been modified by exchange with soil carbonate, which is assumed to have a ratio of 0 per mil, there is a dilution of about 20 percent in the sample, which translates into a reduction of about 2,000 years in the estimated water age for the well 2N/1E-12H2. This nonetheless indicates an old age for water sampled from this relatively shallow well.

The sample from thermal well 2N/1E-12H1 contained the oldest water sampled, with an uncorrected ^{14}C age of about 17,200 years before present, supporting the interpretation of the stable isotope data that the thermal water was recharged during a colder climate. The sample from well 2N/1E-12Q3 was the next oldest (about 6,500 years before present). This well is the deepest well sampled, and is screened from 130 to 790 ft bls. As shown on *figure 37*, well 2N/1E-12Q3 yields water from different zones with different stable isotope values. The sample from 725 ft bls has a similar isotopic signature as well 2N/1E-12H1, indicating that the sample from the lower zone of the well was recharged during the same period as the sample from well 2N/1E-12 H1. The younger apparent age of the well discharge is the result of mixing of younger water from shallower zones in the well. The sample from well 2N/1E-20R1 contained measurable tritium (modern water) and had a ^{14}C age of 2,250 years before present, indicating that this well receives water from zones of differing ages.

Helium-4

The inert nature of Helium, coupled with its distinctive isotopic and solubility characteristics, makes it an ideal tracer in groundwater studies. Helium-4 (^{4}He) concentrations in groundwaters often exceed the expected solubility equilibrium values owing to subsurface production in the aquifer matrix and subsequent release into groundwater (Andrews and Lee, 1979; Andrews and others, 1982; Kulongoski and others, 2003; 2005). Subsurface addition of ^{4}He changes the ^{3}He/^{4}He ratio from that of air-equilibrated water to diagnostic values depending on the lithology of the aquifer, providing a means to identify the He source. Helium-4 accumulates in groundwater from the α-decay of uranium (U) and thorium (Th) series elements within the aquifer material (in situ production; for example, Andrews, 1985; Torgersen, 1980). By measuring the accumulation rate of ^{4}He produced in place, the parent composition of the aquifer rock can be used to determine the travel time of the water through the aquifer.

The amount of helium produced in place that accumulates in groundwater depends upon the radioelement content and porosity of the aquifer and can be quantified using the equation (Andrews and Lee, 1979):

$$^4He_{sol} = \rho \times \Lambda$$
$$\times \left\{1.19 \times 10^{-13}[U] + 2.88 \times 10^{-14}[Th]\right\} \times \frac{(1-\phi)}{\phi} \qquad (5)$$

where

$^4He_{sol}$	is the ^4He solution rate (cm^3 STPg^{-1} H$_2$0 yr^{-1});
[U] and [Th]	are the uranium and thorium concentrations in the aquifer material (ppm);
1.19×10^{-13}	is the ^4He production rate for U (cm^3 STP yr^{-1} µg^{-1} natural U) in equilibrium with its decay products;
2.88×10^{-14}	is the ^4He production rate for Th (cm^3 STP yr^{-1} µg^{-1} natural U) in equilibrium with its decay products;
ρ	is the bulk density of the aquifer material (g cm^{-3});
Λ	is the fraction of helium produced in the rock that is released into the water, assumed to be unity (dimensionless); and
ϕ	is the fractional effective porosity of the aquifer material (dimensionless).

If excess helium in groundwater solely is attributed to production in place, then the helium age of the groundwater can be estimated by dividing the excess ^4He by the solution rate.

Helium-4 concentrations in Big Bear groundwater samples ranged from 3.46 × 10^{-8} to 1,510 × 10^{-8} cm^3 STP g^{-1} H$_2$0. The ^4He concentration was about 22 × 10^{-8} cm^3 STP g^{-1} H$_2$0 in well 2N/1E-12M1, where the presence of tritium (^3H) and the carbon-14 activity indicates modern water, and about 1,510 × 10^{-8} cm^3 STP g^{-1} H$_2$0 in well 2N/1E-12H1 near Pan Hot Springs, where the uncorrected carbon-14 age is 17,200 years old. The correlation between ^4He concentrations and uncorrected carbon-14 ages can be used to conclude that the samples from well 2N/1E-2N1, well 2N/1E14B1, and spring 2N/2E30D1, where only ^4He concentrations were analyzed (less than 10 × 10^{-8} cm^3 STP g^{-1} H$_2$0; *table 16* and *fig. 36*), have a relatively young groundwater age.

Summary of Isotopic Data

The $\delta^{18}O$ and δD composition of groundwater samples collected from the study area are similar to or lighter (more negative) than modern winter precipitation collected on the valley floor at the Big Bear City Airport, indicating that that the predominant source of recharge to Big Bear Valley is winter precipitation. Samples from three of the wells had $\delta^{18}O$ and δD values significantly lighter (more negative) than modern winter precipitation, indicating that the groundwater sampled from these wells was recharged at higher altitudes or cooler conditions than modern precipitation on the valley floor. Tritium and carbon-14 were analyzed in samples from six wells in the valley. Samples collected from two wells with shallow perforations (less than 100 ft bls) contained detectable tritium concentrations, indicating that these wells have received recharge within the past 50 years. Uncorrected carbon-14 groundwater ages (residence times) of the six samples ranged from modern to 17,200 years old. Most of the samples collected for this study were from wells that receive water from different zones or aquifers that contain water of different ages.

Depth-dependent samples collected from well 2N/1E-12Q3, with intermittent perforations from 91 to 790 ft bls, indicated that the stable isotopes of oxygen and hydrogen become lighter with depth. These isotopic data indicate that the groundwater sampled at deeper depths in the well was recharged at higher altitudes than the shallow samples or was recharged during a cooler climate. The sample collected from near the bottom of well (725 ft bls) had a similar temperature and isotopic signature as water sampled from the thermal well at Pan Hot Springs (2N/1E-12H1), indicating that both samples were derived from the same source and were recharged near the end of the last North American glaciation, when it likely was colder and(or) wetter in the study area. Helium-4 data collected from well 2N/1E-12H1 support the great age of the thermal water. The results from this study indicate that isotopic data is useful to help determine the source and age of groundwater in the Big Bear Valley.

Summary and Conclusions

The Big Bear area currently relies on springs on the periphery of the groundwater basin and wells drilled within the basin for its water supply. Because the population in the area has increased in recent years, local water agencies constructed new wells and have studied combining artificial recharge with reclaimed wastewater to help meet demand. To better manage the groundwater resources in Big Bear Valley, there is a need to better understand the geohydrology of the groundwater basin—in particular, the size and shape of the groundwater basin and the quantity, distribution, and age of natural groundwater recharge.

Thickness and Structure of the Groundwater Basin

A gravity survey was used to estimate the thickness of Quaternary alluvial deposits and Tertiary sedimentary deposits that fill the Big Bear groundwater basin (thickness of basin-fill deposits or depth to basement rocks) and to understand the three-dimensional structure (geometry) of the groundwater basin. The gravity field of the study area is complex and mostly reflects lateral variations in the density of the basement rock. The thickness of the groundwater basin was estimated by using an inversion method that permits the inclusion of constraints at points where the thickness of the groundwater basin is known from direct observations of drill holes and geologic maps. The best resolution of interpreted thickness of the groundwater basin was determined to be about ±50 ft and is likely to be greater in areas for which data is poorly constrained. Calculations were made using grid cells 820 ft on a side, so results represent averages of the groundwater-basin thickness over this cell size.

Results indicate that the groundwater basin reaches a maximum thickness of 1,500 to 2,000 ft beneath Big Bear Lake and the area between Big Bear and Baldwin Lakes. The groundwater basin thins to less than 500-ft thick beneath the center and eastern end of Big Bear Lake. The calculated groundwater basin thickness was compared to the measured thickness where wells penetrated the entire aquifer; all the calculated thicknesses were within 50 ft of those measured. The gravity method used for this study does not differentiate between water-bearing and non-water-bearing deposits; therefore, the calculated combined thickness of these deposits cannot be used to estimate the groundwater availability in the groundwater basin independently. Only detailed independent information on the specific thickness and water-bearing properties of these different deposits, which could be provided by borehole data, would permit an accurate estimate of groundwater availability. The thickness map prepared for this study could be used to help identify the location of potential boreholes to investigate the water-bearing properties of the groundwater basin. For example, areas on the map where the

alluvial deposits are identified as having a substantial thickness, and where there is no existing geologic information, could be good locations to drill exploratory boreholes.

Interferometric Synthetic Aperture Radar (InSAR) was used in this study to measure pumping-induced land subsidence and locate structures, such as faults, that can affect groundwater movement. The 16 interferograms developed for this study indicate that small amounts of land-surface deformation occurred in the groundwater basin. In general, land-surface deformation was identified in three areas: (1) the area between Big Bear Lake and Baldwin Lake, (2) the area near the city of Big Bear Lake, and (3) the area near Sugarloaf. In the area between Big Bear and Baldwin Lakes, eleven interferograms indicate land-surface subsidence and five interferograms indicate uplift for various time periods between December 25, 1992, and May 30, 2005. An interferogram for September 25, 1995, through July 21, 1997, showed a maximum of 1.2 in. of subsidence in the Big Bear City Community Services District well field between the eastern end of Big Bear Lake and the western end of Baldwin Lake. Available geologic logs indicate that this area contains layers of silt and clay. Water levels during 1992–2000 did not drop below the lows of 1990 and 1991 and therefore did not exceed the preconsolidation stress during this 8-year period. These data suggest that either the subsidence that occurred during this time was elastic and responding to seasonal water-level fluctuations, or the subsidence was residual and mostly inelastic, responding to the water-level lows of 1990 and 1991 or earlier unknown lows.

In the area near the city of Big Bear Lake, five interferograms showed subsidence (0.4 in. or less), whereas three interferograms showed uplift (0.2 in.) for various time periods between June 18, 1993, and May 30, 2005. It is unclear whether the small amount of deformation in this area is elastic or inelastic because the water-level record before 1993 is incomplete. Small amounts of subsidence (0.6 in. or less) in the Sugarloaf area were identified in five interferograms and uplift (0.4 in. or less) was identified in two interferograms for various time periods between June 18, 1993, and May 30, 2005. The geologic and water-level data suggest that stresses in the Sugarloaf area are elastic, and that the small amounts of deformation in the area are recoverable. The interferograms show a northwest-southeast trending linear feature between Big Bear and Baldwin Lakes, implying the presence of a fault or an abrupt change in lithology, but this is uncertain.

Although the subsidence detected in this study is relatively small, inelastic affirmation could provide information to assist local management activities, such as choices concerning additional wells or artificial recharge projects. If InSAR investigations determined the presence of a fault in the area between Big Bear and Baldwin Lakes, and the fault was confirmed by further investigation, implications with regard to groundwater flow and management could be important.

Groundwater Recharge

Two modeling approaches were undertaken to evaluate the distribution and quantity of groundwater recharge in the Big Bear Lake and Baldwin Lake surface-water drainage basins to provide multiple lines of evidence to support conclusions about recharge. These modeling approaches included developing a monthly water-balance model, referred to as the Basin Characterization Model (BCM) and a daily rainfall-runoff model, referred to as INFILv3. The monthly BCM was based on a regional-scale application calibrated to recharge estimates for multiple basins in the southwestern United States. Model results are useful for bounding water-balance results of more detailed models, evaluating long-term climate conditions, illustrating the mechanisms responsible for recharge in a basin, and comparing recharge and runoff in different basins on a regional scale. The daily INFILv3 model was calibrated to lake-level and volume records available for Big Bear Lake and lake-level records available for Baldwin Lake, and provides a detailed water balance by using daily climate input.

The BCM used monthly precipitation and air-temperature maps for 1971 to 2000 to determine potential recharge for 11 subbasins of the Big Bear Lake and Baldwin Lake surface-water drainage basins. The BCM incorporated GIS coverages of soil, geology, and topographic information, and additional process models to develop spatial distributions of potential evapotranspiration, snow accumulation, and snowmelt. A simple water-balance equation was used with surface properties of soil-water storage and saturated hydraulic conductivity of bedrock to determine the amount of water potentially available for runoff and in-place recharge. The BCM simulated approximately 12,700 acre-ft/yr of potential in-place recharge and approximately 30,800 acre-ft/yr of potential runoff. If 10 percent of the runoff became recharge, approximately 15,800 acre-feet of total recharge would occur the Big Bear study area—about 6,630 acre-ft/yr in the Big Bear Lake surface-water drainage basin and about 9,170 acre-ft/yr in the Baldwin Lake surface-water drainage basin.

The INFILv3 model used a daily time step for simulating the root-zone water balance and an hourly time step for simulating potential evapotranspiration. The daily water balance included precipitation, snow accumulation, sublimation, snowmelt, infiltration into the root zone, evapotranspiration, percolation and redistribution through a multi-layered root zone, water-content change throughout the root-zone, surface-water runoff, and net infiltration through the bottom of the root zone. The INFILv3 model was modified to include a shallow groundwater zone (SGWZ) underlying the root zone, and the processes of lateral seepage from the SGWZ, and recharge through the bottom of the SGWZ.

An initial calibration to the observed lake volumes at Big Bear and Baldwin Lakes was attempted by using a base-case climate model. The base-case model configuration consisted of a preliminary-climate input for INFILv3 using monthly PRISM maps and daily climate data from 35 NCDC stations.

The base-case climate model simulated a total of about 88,200 acre-ft/yr of total precipitation, which was evenly divided between Big Bear and Baldwin Lakes surface-water drainage basins. The initial calibration provided a very good match between simulated and observed lake volumes for Big Bear Lake, but lake volumes were overestimated for Baldwin Lake, and a simultaneous calibration to both lakes using a consistent set of model parameters could not be achieved. The calibration results for Baldwin Lake indicated that INFILv3 base-case climate model simulated too much potential inflow to Baldwin Lake to obtain a satisfactory fit between simulated and measured lake volumes, even when assuming no groundwater inflow to the lake area. A sensitivity analysis indicated that a good calibration for Baldwin Lake was obtained only by reducing all inflows generated by the base-case INFILv3 model (precipitation, surface-water runoff, and groundwater), indicating that the precipitation input for the base-case model was too high for the Baldwin Lake surface-water drainage basin.

To improve the match between measured and spatially-interpolated precipitation in the Baldwin Lake surface-water drainage basin, the preliminary climate input was revised by modifying the PRISM monthly precipitation maps. As with the preliminary climate input, the PRISM monthly precipitation maps were used to define the spatial-interpolation model for precipitation, and the monthly maximum and minimum air temperature for the revised climate input. The modified PRISM monthly precipitation maps were developed using the ratio of recorded average monthly precipitation to PRISM average monthly precipitation at 84 NCDC, RAWS, CIMIS and SBC stations having at least 10 years of record for a given month. The adjusted PRISM data provided a better match to the drier average monthly precipitation measured in the Baldwin Lake drainage basin, resulting in an improved representation of localized rain-shadow effects. The INFILv3 revised climate model calibrated well to the Baldwin Lake record. Calibration results for Big Bear Lake, although not as good as results obtained using the base-case model, were still well within the criteria used to indicate a satisfactory calibration. The revised climate model was selected as most appropriate for estimating spatially and temporally distributed recharge and for evaluating the water balance for the Big Bear study area.

The INFILv3 simulation results obtained with the revised climate model for water years 1950–2005 were used to develop time-averaged, spatially-distributed estimates for all components of the water balance for all subbasins in the Big Bear and Baldwin Lakes surface-water drainage basins. Total precipitation (rain and snow) for both basins was about 73,290 acre-ft/yr using the revised climate model—rain and snow were fairly evenly divided. A total of about 5,480 acre-ft/yr was simulated as recharge in the Big Bear study area, with about 2,800 acre-ft/yr in the Big Bear Lake surface-water drainage basin and about 2,680 acre-ft/yr in the Baldwin Lake surface-water drainage basin. The simulated total recharge was about 11 percent of the total basin precipitation for the land areas upstream of the lakes.

The INFILv3 revised climate model simulated average recharge for water-years 1950–2005 (about 5,480) is about 35 percent of the value estimate using BCM for water years 1971–2000 (about 15,800 acre-ft/yr). The main reasons for this difference are that the precipitation simulated by the INFILv3 revised climate model (about 73,290 acre-ft/yr) is about 20 percent lower than simulated by the BCM (about 92,050 acre-ft/yr) and that the INFILv3 model simulates lateral seepage, which results in a greater amount of evapotranspiration compared to the BCM. The daily runoff simulated by the INFILv3 model allowed for a comparison of simulated to measured lake levels and volumes for Big Bear and Baldwin Lakes. This comparison, indicated that the PRISM distribution of precipitation, used in both the BCM and base-case INFILv3 models, overestimated the precipitation in the Baldwin surface-water drainage basin.

The modeling approaches used for this study indicate that the groundwater recharge in the Big Bear and Baldwin Lakes surface-water drainage basins ranges from about 5,500 to 16,000 acre-ft/yr, with most of the recharge occurring in the mountains surrounding the valley floor. However, if it is assumed that all of the simulated surface-water outflow and groundwater recharge in the Big Bear and Baldwin Lakes surface-water drainage basins discharges into the downstream lakes, then the lower range of these values probably is more representative of actual recharge, as indicated by the good match between the INFILv3 revised climate model simulated and measured lake volumes. A fully coupled surface-water/groundwater-flow model could be used to better refine the recharge estimate.

Source and Age of Groundwater

One spring and eight wells were sampled and analyzed for chemical and isotopic data in 2005 and 2006 in the Big Bear Valley to determine if isotopic techniques could be used to assess the sources and ages of groundwater in the valley. Selected samples were analyzed for stable isotopes of hydrogen and oxygen ($\delta^{18}O$ and δD), tritium, carbon-13 (^{13}C), carbon-14 (^{14}C), and helium (^{4}He). A velocity log, temperature log, and depth-dependent water-quality data were collected from well 2N/1E-12Q3 to determine if the isotopic characteristics of the groundwater basin varied with depth.

The $\delta^{18}O$ and δD composition of groundwater samples collected from the study area are similar to or lighter (more negative) than modern winter precipitation collected at the Big Bear City Airport, on the valley floor, suggesting that that the predominant source of recharge to the Big Bear Valley is winter precipitation. Samples from three of the wells had $\delta^{18}O$ and δD values significantly lighter (more negative) than modern winter precipitation, indicating that the groundwater sampled from these wells derived from higher altitudes or cooler conditions than modern precipitation on the valley floor. Tritium and carbon-14 were analyzed in samples from six wells in the valley. Samples collected from two wells with shallow perforations (less than 100 ft bls) contained detectable tritium concentrations, indicating that these wells have received recharge within the past 50 years. Uncorrected carbon-14 groundwater ages (residence times) of the six samples ranged from modern to 17,200 years old. Most of the samples collected for this study were from wells that receive water from different zones or aquifers that contain water of different ages.

Depth-dependent samples collected from a well 2N/1E-12Q3, with intermittent perforations from 91 to 790 ft bls, indicated that the stable isotopes of oxygen and hydrogen become lighter with depth. These isotopic data indicate that the groundwater sampled at deeper depths in the well was recharged at higher altitudes than the shallow samples or was recharged during a colder climate. The sample collected from near the bottom of well (725 ft bls) had a similar temperature and isotopic signature as water sampled from the thermal well at Pan Hot Springs (2N/1E-12H1), implying that both samples were derived from the same source and were recharged near the end of the last North American glaciation, when it likely was colder and(or) wetter in the study area. Helium-4 data collected from well 2N/1E-12H1 supports the great age of the thermal water. The results from this study indicate that isotopic data can be useful to help determine the source and age of groundwater in the Big Bear study area.

Information Useful to Groundwater Management

The results of these investigations provide an understanding of the lateral and vertical extent of the groundwater basin, its general structural features and properties, the spatial and temporal distribution of groundwater recharge, and the dominant processes responsible for the recharge. Information presented in this report, including estimates of the thickness of alluvial and sedimentary deposits within the groundwater basin, recharge locations and amounts, and the age of groundwater, is intended to help decision-makers evaluate groundwater storage and possible locations for future wells. This information does not include detailed subsurface properties necessary to quantify the availability of water affected by localized pumping or recharge activities; this would require more detailed subsurface characterization of the lithologic units that compose the groundwater basin. These studies do, however, provide a base-line characterization of the hydrogeologic framework and rainfall-runoff characteristics, and could be directly applied to the development of a fully coupled surface-water/groundwater model. A coupled surface-water/groundwater model could be calibrated to the water-level data, subsidence evidence, and available isotopic and chemical data, and this would greatly improve the understanding of the potential hydrologic effects of different water-management alternatives on groundwater levels and movement in the Big Bear Valley groundwater basin. This study provides an example of using multiple approaches to constrain the estimates of groundwater availability in a data-poor basin, and provides insights into possible characterization approaches in other basins.

References Cited

Alley, W.M., Reilly, T.E., and Franke, O.L., 1999, Sustainability of Groundwater Resources: U.S. Geological Survey Circular 1186, 79 p.

Anderson, E. A., 1976, A point energy and mass balance model of a snow cover: Silver Spring, Md., U.S. National Oceanographic and Atmospheric Administration (NOAA) Technical Report NWS 19.

Andrews, J.N., 1985, The isotopic composition of radiogenic helium and its use to study ground water movement in confined aquifers: Chemical Geology, v. 49, p. 339–351.

Andrews, J.N., Giles, I.S., Kay, R.L.F., Lee, D.J., Osmond, J.K., Cowart, J.B., Fritz, P., Barker, J.F., and Gale, J., 1982, Radioelements, radiogenic helium and age relationships for ground waters from the granites at Stripa, Sweden: Geochemica et Cosmochimica, v. 46, p. 1533–1543.

Andrews, J.N., Goldbrunner, J.E., Darling, W.,G., Hooker, P.J., Wilson, G.B., Youngman, M.J., Eichinger, L., Rauert, W., Stichler, W., 1985, A radiochemical, hydrochemical and dissolved gas study of ground waters in the Molasse basin of Upper Austria: Earth and Planetary Science Letters, v. 73, p. 317–332.

Andrews, J.N., and Lee, D.J., 1979, Inert gases in ground water from the Bunter Sandstone of England as indicators of age and paleoclimatic trends: Journal of Hydrology, v. 41, p. 233–252.

Bicknell, B.R., Imhoff, J.C., Kittle, J.L., Jr., Donigian, A.S., Jr., and Johanson, R.C., 1997, Hydrological Simulation Program-FORTRAN, User's manual for Version 11: Athens, Ga., U.S. Environmental Protection Agency Report no. EPA/600/R-97/080, 775 p.

Brinkmann R., Münnich K.O. and Vogel J.C., 1959, 14C Altersbestimmung von Grundwasser, Naturwissenschaften 46, p. 10–12.

California Department of Water Resources, 2003, California's groundwater: Bulletin 118, update 2003: Sacramento, Calif., State of California, The Resources Agency, Department of Water Resources, accessed on July 1, 2004, at URL *http://www.dpla2.water.ca.gov/publications/groundwater/ bulletin118/basins/pdfs_desc/8-9.pdf*

Chow, V.T., 1964, Geological map of the San Gorgonio Mountain Quadrangle, San Bernardino and Riverside Counties, California, Map I: U.S. Geological Survey.

Coplen, T.B., Wildman, J.D., and Chen, J., 1991, Improvements in the gaseous hydrogen-water equilibrium technique for hydrogen isotope analysis: Analytical Chemistry, v. 63, p. 910–912.

Coplen, T.B., 1994, Reporting of stable hydrogen, carbon, and oxygen isotopic abundances: Pure and Applied Chemistry, v. 66, p. 273–276.

Craig, Harmon, 1961, Isotope variations in meteoric water: Science, v. 133, p. 1702–1703.

Crippen, J.R., 1965, Natural water loss and recoverable water in mountain basins of southern California: U.S. Geological Survey Professional Paper 417-E, p. E1–E24.

Daly, C., Gibson, W.P., Doggett, M., Smith, J., and Taylor, G., 2004, Up-to-date monthly climate maps for the conterminous United States: Procedures of the 14th American Meteorological Society Conference on Applied Climatology, 84th AMS Annual Meeting Combined Preprints, American Meteorological Society, Seattle, Washington, January 13–16, 2004, Paper P5.1, CD-ROM.

Daly, C., Neilson, R.P., and Phillips, D.L., 1994, A statistical-topographic model for mapping climatological precipitation over mountain terrain: Journal of Applied Meteorology, v. 33, no. 2, p. 140–158.

Dansgaard, W., 1964, Stable isotopes in precipitation: Tellus, v. 16, p. 436–469.

Donahue, D.J., Linick, T.W., and Jull, A.J.T., 1990, Ratio and background corrections for accelerator mass spectrometry radiocarbon measurements: Radiocarbon, v. 32, p. 135–142.

EarthInfo, Inc., 2006, NCDC Summary of the Day, West: Boulder, Colo., Earth Info, Inc., CD ROM WestEpstein, S., and Mayeda, T., 1953, Variation of O-18 content of water from natural sources: Geochimica Cosmochimica Acta, v. 4, p. 213–224.

Faires, L.M., 1993, Methods of analysis by the U.S. Geological Survey National Water Quality Laboratory–Determination of metals in water by inductively coupled plasma-mass spectrometry: U.S. Geological Survey Open-File Report 92-634, 28 p.

Fishman, M.J., 1993, Methods of analysis by the U.S. Geological Survey National Water Quality Laboratory–Determination of inorganic and organic constituents in water and fluvial sediments: U.S. Geological Survey Open-File Report 93-125, 217 p.

Fishman, M.J., and Friedman, L.C., 1989, Methods for determination of inorganic substances in water and fluvial sediments: U.S. Geological Survey Techniques of Water-Resources Investigations, book 5, chap. A1, 545 p.

Flint, A.L., and Childs, S.W., 1987, Calculation of solar radiation in mountainous terrain: Journal of Agricultural and Forest Meteorology, v. 40, p. 233–249.

Flint, A.L., Flint, L.E., Hevesi, J.A., D'Agnese, F.A., and Faunt, C.C., 2000, Estimation of regional recharge and travel time through the unsaturated zone in arid climates, *in* Faybishenko, B., Witherspoon, P., and Benson, S. (eds.), Dynamics of Fluids in Fractured Rock: Washington, D.C., Geophysical Monograph 122, American Geophysical Union, p. 115–128.

Flint, A.L., Flint, L.E., Hevesi, J.A., and Blainey, J.M., 2004, Fundamental concepts of recharge in the Desert Southwest: a regional modeling perspective, in Hogan, J.F., Phillips, F.M., and Scanlon, B.R., eds., Groundwater Recharge in a Desert Environment: The Southwestern United States: , Washington, D.C., Water Science and Applications Series, v. 9, American Geophysical Union, p. 159–184.

Flint, A.L., Flint, L.E., and Masbruch, M.D., 2011, Input, calibration, uncertainty, and limitations of the Basin Characterization Model: Appendix 3 of Conceptual Model of the great Basin Carbonate and Alluvial Aquifer System (eds. V.M. Heilweil and L.E. Brooks), U.S. Geological Survey Scientific Investigations Report 2010–5193.

Flint, L.E. and Flint, A.L., 1995, Shallow infiltration processes at Yucca Mountain– Neutron logging data 1984-93: U.S. Geological Survey Water-Resources Investigations Report 95-4035, 46 p.

Flint, A.L., and Flint, L.E., 2007a, Application of the basin characterization model to estimate in-place recharge and runoff potential in the Basin and Range carbonate-rock aquifer system, White Pine County, Nevada, and adjacent areas in Nevada and Utah: U.S. Geological Survey Scientific Investigations Report 2007–5099, 20 p.

Flint, L.E., and Flint, A.L., 2007b, Regional analysis of runoff and groundwater recharge: Chapter B in Stonestrom, D.A., Constantz, J., Ferre, P.A., and Leake, S.A., eds., Groundwater Recharge in the Arid and Semi-arid Southwestern USA, U.S. Geological Survey Professional Paper 1703, p. 29–59.

Friedman, I., Smith, G.I., Gleason, J.D., Warden, A., and Harris, J.M., 1992, Stable isotope composition of waters in Southeastern California: 1. Modern Precipitation: Journal of Geophysical Research, v. 97, issue D5, p. 5795–5812.

Galloway, D.L., Jones, D.R., and Ingebritsen, S.E., 1999, Land subsidence in the United States: U.S. Geological Survey Circular 1182, 175 p.

Garbarino, J.R., 1999, Methods of analysis by the U.S. Geological Survey National Water Quality Laboratory – determination of dissolved arsenic, boron, lithium, selenium, strontium, thallium, and vanadium using inductively coupled plasma-mass spectrometry: U.S. Geological Survey Open-File Report 99-093, 31 p.

Garbarino, J.R., Kanagy, L.K., and Cree, M.E., 2006, Determination of elements in natural-water, biota, sediment and soil samples using collision/reaction cell inductively coupled plasma-mass spectrometry: U.S. Geological Survey Techniques and Methods, book 5, chap. B1, 88 p.

GeoScience Support Services, Inc., 1992, Draft re-evaluation of maximum perennial yield Big Bear ground water basins, San Bernardino County, California: Prepared for City of Big Bear Lake, Department of Water And Power, June 15, 1992.

GeoScience Support Services, Inc., 1999, Re-evaluation of the maximum perennial yield in the Baldwin Lake Watershed: Report prepared for the Big Bear City Community Services District, July 13, 1999, 682 p.

GeoScience Support Services, Inc., 2001, Re-evaluation of the maximum perennial yield Big Bear Lake watershed and a portion of Baldwin Lake Watershed: Prepared for City of Big Bear Lake Department of Water and Power, August 24, 2001.

GeoScience Support Services, Inc., 2003, Focused geohydrologic evaluation of the maximum perennial yield of the North Shore and Grout Creek hydrologic subunit tributary subareas: Report prepared for the City of Big Bear Lake Department of Water and Power, Dec. 2, 2003, 67 p.

GeoScience Support Services, Inc., 2004a, Geohydrologic evaluation of the maximum pereinnial yield of the Lake William Area, Baldwin Lake Watershed: Report prepared fpr the City of Big Bear Lake Department of Water and Power, Jan, 27, 2004.

GeoScience Support Services, Inc., 2004b, Resuls of drilling and testing—Big Bear Valley ground water exploration program 2003: Report prepared for the City of Big Bear Lake Department of Water and Power, March 5, 2004.

GeoScience Support Services, Inc., 2005a, Resuls of drilling, construction, development and testing-McAlister production well: Report prepared for the City of Big Bear Lake Department of Water and Power, Feb. 24, 2005.

GeoScience Support Services, Inc., 2005b, Results of drilling, construction, development and testing—Canvasback production well: Report prepared for the City of Big Bear Lake Department of Water and Power, Feb. 24, 2005.

GeoScience Support Services, Inc., 2006, Technical memorandum perennial yield update for the City of Big Bear Lake Department of Water and Power Service area : Report prepared for the City of Big Bear Lake Department of Water and Power, Feb. 2, 2006, 26 p.

Gleason, J.D., Veronda, G., Smith, G.I., Friedman, I., and Martin, P., 1994, Deuterium content of water from wells and perennial springs, southeastern California: U.S. Geological Survey Hydrologic Investigations Atlas HA-727.

Gonfiantini, R., 1978, Standards for stable isotope measurements in natural compounds: Nature 271, p. 534–536.

Hevesi, J.A., Flint, A.L., and Flint, L.E., 2003, Simulation of net infiltration using a distributed parameter watershed model for the Death Valley regional flow system, Nevada and California: U.S. Geological Survey Water-Resources Investigation Report 03–4090, 104 p.

Ingraham, N.L., and Taylor, B.E., 1991, Light stable isotope systematics of large-scale hydrologic regimes in California and Nevada: Water Resources Research, v. 27, p. 77–90.

Izbicki, J.A., 2002, Geologic and hydrologic controls on the movement of water through a thick, heterogeneous unsaturated zone underlying an intermittent stream in the western Mojave Desert, southern California: Water Resources Research, v. 38, p. 1–14.

Izbicki, J.A., Pimentel, M., Leddy, M., and Bergamaschi, B., 2004, Microbial and dissolved organic carbon characterization of stormflow in the Santa Ana River at Imperial Highway, southern California: U.S. Geological Survey Scientific Investigations Report 2004–5116, 71p., also available at *http://water.usgs.gov/pubs/sir/2004/5116*.

Jachens, R.C., and Moring, B.C., 1990, Maps of thickness of Cenozoic deposits and isostatic residual gravity over basement in Nevada: U.S. Geological Survey Open-File Report 90-404.

Jachens, R.C., and Roberts, C.W., 1981, Documentation of a FORTRAN program, 'isocomp', for computing isostatic residual gravity: U.S. Geological Survey Open-File Report 81-574, 26 p.

Jennings, C.W., 1977, Geologic map of California: California Division of Mines and Geology Geologic Data Map no. 2, scale 1:750,000.

Jull, A.J.T., Burr, G.S., McHargue, L.R., Lange, T.E., Lifton, N.A., Beck, J.W., Donahue, D., and Lal, D., 2004, New frontiers in dating of geological, paleoclimatic and anthropological applications using accelerator mass spectrometric measurements of ^{14}C and ^{10}Be in diverse samples: Global & Planetary Change, v. 41, p. 309–323.

Kalin, R.M., 2000, Radiocarbon dating of groundwater systems, *in* Cook, P.G., and Herczeg, A.L., eds., Environmental tracers in subsurface hydrology, chap. 4: Boston, Kluwer Academic Publishers, p. 111–144.

Karki, P., Kivioja, L., and Heiskanen, W.A., 1961, Topographic isostatic reduction maps for the world for the Hayford Zones 18-1, Airy-Heiskanen System, T=30 km: Isostatic Institute of the International Association of Geodesy, no. 35, 5 p., 20 pl.

Kulongoski, J.T. and Hilton, D.R., 2002. A quadrupole-based mass spectrometric system for the determination of noble gas abundances in fluids. Geochem. Geophys. Geosyst. 3, U1– U10.

Kulongoski, J.T., Hilton, D.R., Izbicki, J.A., 2003, Helium isotope studies in the Mojave Desert, California: Implications for groundwater chronology and regional seismicity: Chemical Geology, v. 202, no. 1-2, p. 95–113.

Kulongoski, J.T., Hilton, D.R., Izbicki, J.A., 2005, Source and movement of helium in the eastern Morongo ground water Basin: The influence of regional tectonics on crustal and mantle helium fluxes: Geochimica et Cosmochimica Acta, v. 69, no. 15, p. 3857–3872.

Leake, S.A., and Prudic, D.E., 1991, Documentation of a computer program to simulate aquifer-system compaction using the modular finite-difference groundwater flow model: U.S. Geological Survey Techniques of Water-Resources Investigation Report, book 6, chap A2, 68 p.

Leavesley, G.H., Lichty, R.W., Troutman, B.M., and Saindon, L.G., 1983, Precipitation-Runoff Modeling System: User's Manual: U.S. Geological Survey Water-Resources Investigations 83-4238, 207 p.

LeRoy Crandall and Associates, 1987, Re-evaluation of sustained ground water yields, Big Bear Lake Watershed, San Bernardino County, California: Prepared for the City of Big Bear Lake.

Lundquist, J.D., and Flint, A.L., 2006, 2004 Onset of snowmelt and streamflow: How shading and the solar equinox may affect spring runoff timing in a warmer world: Journal of Hydroclimatology, v. 7, p. 1199–1217.

Mabey, D.R., 1960, Gravity survey of the western Mojave Desert, California: U.S. Geological Survey Professional Paper 316-D, p. 51–73.

Maidment, D.R., 1993, Handbook of Hydrology: McGraw-Hill, Inc., p. 7.5.

Maidment, D.R. (ed), 2002, Arc Hydro:GIS for Water Resources: Redlands, California, ESRI Press.

McLain, B.J., 1993, Methods of analysis by the U.S. Geological Survey National Water Quality Laboratory; determination of chromium in water by graphite furnace atomic absorption spectrophotometry: U.S. Geological Survey Open-File Report 93-449, 16 p.

Meinzer, O.E., 1928, Compressibility and elasticity of artesian aquifers: Economic Geology, v. 23, no. 3, p. 263–291.

Michel, R.L, 1976, Tritium inventories of the world oceans and their implications: Nature, v. 263, p. 103–106.

Miller, F.K., 2004, Preliminary geologic map of the Big Bear City 7.5′ quadrangle, San Bernardino County, California: U.S. Geological Survey Open-File Report 2004–1193, scale 1:24,000.

Miller, F.K., Matti, J.C., Brown, H.J., and Powell, R.E., 2001, Geologic map of the Fawnskin 7.5′ quadrangle, San Bernardino County, California: U.S. Geological Survey Open-File Report 98-579, scale 1:24,000.

Mook, W.G., 1980, The dissolution-exchange model for dating of groundwater with ^{14}C, *in* Fritz, P. and Fontes, J.C., eds., Handbook of Environmental Isotopes Geochemistry, v. 1: Amsterdam, Elsevier, p. 50–74.

Nalder, I.A., and Wein, R.W., 1998, Spatial interpolation of climatic Normals: Test of a new method in the Canadian boreal forest: Agricultural and Forest Meteorology, v. 92, no. 4, p. 211–225.

Nash, J.E., and Sutcliffe, J.V., 1970, River flow forecasting through conceptual models part I—A discussion of principals: Journal of Hydrology, v. 10, no. 3, p. 282–290.

Newhouse, M.W., Izbicki, J.A., and Smith, G.A., 2005, Comparison of velocity-log data collected using impeller and electromagnetic flowmeters: Ground Water, v. 43, no. 3, p. 434–438.

Nishikawa, Tracy, Izbicki, J.A., Hevesi, J.A., Stamos, C.L., and Martin, Peter, 2004, Evaluation of geohydrologic framework, recharge estimates, and groundwater flow of the Joshua Tree area, San Bernardino County, California: U.S. Geological Survey Scientific Investigations Report 2004–5267, 127 p.

Ostlund, H.G., and Werner, E., 1962, The electrolytic enrichment of tritium and deuterium for natural tritium measurements, in Tritium in the Physical and Biological Sciences, v. I: International Atomic Energy Agency, Vienna, p. 95–104.

Patton, C.J., and Kryskalla, J.R., 2003, Methods of Analysis by the U.S. Geological Survey National Water Quality Laboratory—Evaluation of alkaline persulfate digestion as an alternative to Kjeldahl digestion for determination of total and dissolved nitrogen and phosphorus in water: U.S. Geological Survey Water-Resources Investigations Report 03–4174, 33 p.

Patton, C.J., and Truitt, E.P., 1992, Methods of analysis by the U.S. Geological Survey National Water Quality Laboratory—Determination of total phosphorus by a Kjeldahl digestion method and an automated colorimetric finish that includes dialysis: U.S. Geological Survey Open-File Report 92-146, 39 p.

Patton, C.J., amd Truitt, E.P., 2000, Methods of analysis by the U.S. Geological Survey National Water Quality Laboratory—Determination of ammonium plus organic nitrogen by a Kjeldahl digestion method and an automated photometric finish that includes digest cleanup by gas diffusion: U.S. Geological Survey Open-File Report 00-170, 31 p.

Pavelko, M.T., 2004, Estimates of hydraulic properties from a one-dimensional numerical model of vertical aquifer-system deformation, Lorenzi Site, Las Vegas, Nevada: U.S. Geological Survey Water-Resources Investigations Report 03-4083, 35 p.

Plouff, D., 1992, Bouguer gravity anomaly and isostatic residual gravity maps of the Reno 1° by 2° quadrangle, Nevada and California: U.S. Geological Survey Miscellaneous Field Studies Map MF-2154-E.

Poland, J.F., ed., 1984, Guidebook to studies of land subsidence due to groundwater withdrawal, in UNESCO Studies and Reports in Hydrology: Paris, France, United Nations Educational, Scientific, and Cultural Organization, v. 40, 305 p., 5 appendixes.

Priestley, C.H.B., and Taylor, R.J., 1972, On the assessment of surface heat flux and evaporation using large-scale parameters: Manual Weather Review, v. 100, p. 81–92.

Rewis, D.L., Christensen, A.H., Matti, J., Hevesi, J.A., Nishikawa, T., and Martin, P., 2006, Geology, groundwater hydrology, geochemistry, and groundwater simulation of the Beaumont and Banning storage units, San Gorgonio Pass Area, Riverside County, California: U. S. Geological Survey Scientific Investigations Report 2006–5026,191 p.

Riley, F.S., 1998, Mechanics of aquifer systems—The scientific legacy of Joseph F. Poland, *in* Borchers, J., ed., Land subsidence—Case Studies and Current Research: Proceedings of the Dr. Joseph F. Poland Symposium on Land subsidence, Association of Engineering Geologists Special Publication 8, p. 13–227.

Roberts, C.W., Jachens, R.C., and Oliver, H.W., 1990, Isostatic residual gravity map of California and offshore southern California: California Division of Mines and Geology California Geologic Data Map Series Map 7, scale 1:750,000.

Roscoe-Moss Co., 1990, Handbook of Ground Water Development: John Wiley & Sons.

Rose, T.P., Davisson, M.L., and Criss, R.E., 1996, Isotope hydrology of voluminous cold springs in fractured rock from an active volcanic region, northeastern California: Journal of Hydrology, v. 179, p. 207–236.

Shamir, E., and Georgakakos, K.P., 2005, Distributed snow accumulation and ablation modeling in the American River Basin: Advances in Water Resources, v. 29, no. 4, p. 558–570

Shuttleworth, W.J., 1993, Evaporation *in* Maidment, D.R., ed., Handbook of Hydrology, chap. 4: McGraw-Hill, Inc., p. 4.24.

Smith, G.I., Friedman, I., Gleason, J.D., and Warden, A., 1992, Stable isotope composition of waters in southeastern California: 2. Groundwaters and their relation to modern precipitation: Journal of Geophysical Research, v. 97, p. 5813–5823.

Sneed, Michelle, and Galloway, D.L., 2000, Aquifer-system compaction and land subsidence: measurements, analyses, and simulations—the Holly Site, Edwards Air Force Base, Antelope Valley, California: U.S. Geological Survey Water-Resources Investigations Report 00-4015, 65 p.

Struzeski, T.M., DeGiacomo, W.J., and Zayhowski, E.J., 1996, Methods of analysis by the U.S. Geological Survey National Water Quality Laboratory—Determination of dissolved aluminum and boron in water by inductively coupled plasma-atomic emission spectrometry: U.S. Geological Survey Open-File Report 96-149, 17 p.

Terzaghi, Karl, 1925, Principles of soil mechanics, IV—Settlement and consolidation of clay: Engineering News-Record, v. 95, no. 3, p. 874.

Thatcher, L.L., Janzer, V.J., and Edwards, K.W., 1977, Methods for the determination of radioactive substances in water: U.S. Geological Survey Techniques of Water-Resources Investigations, chap. A5, 95 p.

Timme, P.J., 1995, National Water Quality Laboratory 1995 services catalog: U.S. Geological Survey Open-File Report 95-352, 120 p.

Todd, D.K., 1980, Groundwater Hydrology: Second Edition, John Wiley & Sons.

Torgersen, T., 1980. Controls on pore-fluid concentrations of ^{4}He and ^{222}Rn and the calculation of ^{4}He/^{222}Rn ages: Journal of Geochemical Exploration, v. 13, p. 57–75.

U.S. Department of Agriculture 1994, State Soil Geographic (STATSGO) Data Base–Data use information: Natural Resource Conservation Service, Miscellaneous Publication no. 1492: Soil Survey Staff, Natural Resources Conservation Service, United States Department of Agriculture. U.S. General Soil Map (STATSGO) for California, accessed 12/12/2001 at *http://soildatamart.nrcs.usda.gov*

U.S. Geological Survey, 2006, National field manual for the collection of water-quality data: U.S. Geological Survey Techniques of Water-Resources Investigations, book 9, chaps. A1–A9, accessed on July 7, 2006, at *http://pubs.water.usgs.gov/twri9A*

U.S. Geological Survey, 2008, Documentation of computer program INFIL3.0—A distributed-parameter watershed model to estimate net infiltration below the root zone: U.S. Geological Survey Scientific Investigations Report 2008–5006, 98 p.

Vogel J.C. and Ehhalt D.H., 1963, The use of carbon isotopes in groundwater studies, Radioisotopes in Hydrology, International Atomic Energy Agency, Vienna, p. 225–240.